ENVIRONMENTAL POLICY AND SOCIETAL AIMS

STUDIES IN ECOLOGICAL ECONOMICS

VOLUME 2

Environmental Policy and Societal Aims

edited by

DENIS REQUIER-DESJARDINS

Centre d'Economie et d'Ethique pour l'Environnement et le Développement,
Université de Versailles-St. Quentin, France

CLIVE SPASH

Cambridge Research for the Environment,
Department of Land Economy,
University of Cambridge,
Cambridge, United Kingdom

and

JAN VAN DER STRAATEN

Department of Leisure Studies,
Tilburg University,
European Centre for Nature Conservation,
Tilburg, The Netherlands

SPRINGER SCIENCE+BUSINESS MEDIA, B.V.

A C.I.P. Catalogue record for this book is available from the Library of Congress.

ISBN 978-0-7923-6040-7 ISBN 978-94-011-4521-3 (eBook)
DOI 10.1007/978-94-011-4521-3

Originally published by Kluwer Academic Publishers in 1999

PREFACE

Denis Requier-Desjardins, Clive L. Spash and Jan van der Straaten

Many books and articles have been published about the ecological economic dimension of various environmental problems. In most cases, authors pay particular attention to approaches which find their origin in economic theory. Sustainable development and environmental improvement are often regarded as of value in themselves as given phenomena and as a result the policy process by which these goals are to be achieved fails to receive full attention. Our belief is that ecological economics should be addressing the institutional and policy making aspects of environmental problems and so covering a broad socio-economic research agenda which differentiates it from mainstream economic approaches. This can be compatible with research conducted by environmental economists or natural scientists but goes beyond the limitations of a positivist approach, is open to multiple perspectives on the same issue, sees conflict resolution as a social process and accepts the need for research addressing political economy.

In this book, the aims of society itself are given a high profile. We are convinced that an in-depth analysis of the policy process is necessary to understand the pitfalls and barriers which society will confront in the process of sustainable development. This means the chapters of this book discuss topics which are commonly regarded as more or less exogenous to the mainstream definition of economic process. The book starts by dealing with issues related to local development and environmental policy. Environmental policy is seen to be more than a national or international concern, and in particular modern environmental concepts, such as sustaining robust ecosystems, are recognised as having numerous implications for local development. In the following chapters the relationship between environmental policy and other policy goals is discussed. Next, attention is given to the process of technological change, and, finally, environmental policy tools and the process of decision-making are examined.

All the papers in this book were presented at the Inaugural Conference of the European branch of the International Society for Ecological Economics entitled 'Ecology, Society, Economy, in Pursuit of Sustainable Development' held at the University of Versailles, St Quentin-en-Yvelines, France, 23-25 May 1996. The conference was a great success with a wide representation of individuals and institutions from across Europe. There were many interesting presentations at this conference, enabling us to compose a collection of articles which address the topic of this book in a variety of ways. In particular, the chapters of this book have been chosen so as to communicate areas and topics which we believe will be of importance to future research.

The editors of this book commented in detail on the first drafts of the papers, which gave authors the opportunity to revise and improve their contributions. The process of feedback and revision with multiple authors in numerous countries made the project of producing this book more complex than most edited volumes. The secretariat of the European Society for Ecological Economics, located at the Centre d'Economie et d'Ethique pour l'Environnement et le Développement at the University of Versailles, St.Quentin-en-Yvelines, gave substantial support to achieving the final product. In particular the work of Sarah Dwyer and Miguel Lopez deserves special mention for production and editorial assistance.

We hope that this book will stimulate the discussion between policy-makers on the one hand, and ecological economists on the other. Such debate is necessary if we are to clarify the specific areas of research which need to be explored further in order to facilitate goals such as environmental improvement, ecosystem maintenance and sustainable development.

Denis Requier-Desjardins
Clive L. Spash
Jan van der Straaten
May 1998

CONTENTS

1 ENVIRONMENTAL POLICY AND SOCIETAL AIMS

Denis Requier-Desjardins, Clive L. Spash and Jan van der Straaten

Introduction

In a democratic society environmental policies need to be related to the full range of objectives to be held important by that society. Without understanding the relevance of this relationship we might expect, for example, that strong resistance from polluters would be the normal situation in all cases where authorities have the intention of implementing policies involving strict environmental rules. This resistance would be successful in many instances, especially when there is weak or poorly co-ordinated support for such strict rules. The successful implementation of legal restrictions, which can be observed in many democratic societies, implies that environmental policy can be established while reinforcing and supporting a range of societal aims. However, the conclusion that the relationship between environmental policies and societal aims is therefore well defined does not follow from this observation.

In part, the relationship is difficult to clarify because the processes leading to the implementation of environmental policies are different from those forming and changing public opinion. In modern democratic countries the formulation and implementation of environmental policies is the initiative of public authorities; in most European cases the Minister of the Environment. Usually, the Minister will avoid taking any initiative without consulting with other ministers, in particular those in charge of the polluting sectors such as agriculture, energy production and transportation. Although there are recognisable differences between countries with respect to the level to which polluting sectors, as power groups, can influence the process of policy formulation, the authorities in democratic countries do tend to discuss the measures to be taken with the polluters prior to their implementation.

The outcome of this process is uncertain for several reasons. First, polluters have different degrees of economic importance in different countries. Thus, for example, the role of intensive farming in the Netherlands in complete contrast with that in Austria or Denmark. Second, certain environmental issues attract more attention in one country than they do in another. Acidic deposition or nuclear contamination due to electric power generation, for example, are regarded differently in Germany than in France or Switzerland. This poses particular problems in the context of the European Union because one aim of that institution is that all member states have the same environmental policies. In this case, the desire for equity in policy terms can be expected to result in international conflict.

While environmental agencies, as branches of government, may take the lead on policy initiation the formation of specific policy is a complex interaction between various groups. During the process of environmental policy design and articulation, discussions take place within the existing democratic institutions. In principal, these institutions allow the established political parties and those groups and individuals with a vested interest in the policy to have their say. The civil servants of the ministries involved write documents to advise the ministers and for wider dissemination. The media react to the documents and play on their perception of the issues of the day, and in doing so they also influence the outcome of the debate. This process of debate and articulation is where sectors of society express their concerns about conflicting societal aims.

An important aspect of this process is that, whenever and wherever environmental policies are implemented there is a change in the rights operating over the environment. Environmental policies, to date, generally aim to restrict, in one way or another, the unlimited use of the environment by polluters (be they industries or individual citizens). In the course of time, the evidence has accumulated to convince many people that the environment is no longer able to absorb the ever increasing level and range of human activities impacting upon ecosystem functions, whether that intervention be in the form of emitting pollution or manipulating genes. New, strict environmental rules are needed to regulate future human interactions with the environment. This can be done through legislation or through the use of economic instruments such as fees and taxes. Thus, in general, the process of environmental policy design and articulation tends to be characterised by a top-down approach. If the environmental problems are recognised by the authorities they will start a process of devising new environmental policies which then need to be considered by all economic actors.

Societal aims are related to this process, but they are at the same time independent of it. In general, the processes changing societal aims are more vague and less well defined than the political process itself. Norms and values, in particular those dealing with environmental issues, can change quite rapidly. Any discussion concerning changes in these values tend to occur more implicitly within society. Opinions result from confronting problems found in society itself in every day life. Many actors influence the debate, such as labour unions, environmental groups, political parties, nature protection authorities, journalists, scientists, etc. Furthermore, the debate extends beyond environmental issues. In many cases, environmental problems are connected with other issues such as the distribution of income and power, the relative position of developing countries, human rights, racism, sexual equality and respect for non-humans. From this complexity, full of arguments and opinions, a certain general level of societal aims will result.

There are also certain apparently autonomous elements in the process which may be overlooked. In this regard, the development of technology is only partially controlled by the political process. Technological possibilities are often seen as value-neutral; they occur and often tend to be accepted with little fuss (perhaps with a grudging fatality). This means that people are inclined to incorporate new technologies into their normal everyday lives, even when they are associated with serious environmental problems. For example, a debate on restricting the high level of pollution and energy use of modern modes of transportation such as aeroplanes and automobiles often makes little progress because these technologies are considered a 'normal' part of modern life. This can be seen in the extent to which the positive rejection or voluntary absence from an individual's life style of the car is regarded as abnormal.

Another complication is the problem of scale and the distribution of environmental impacts. Certain environmental problems, such as the exact location of a municipal waste dump, are only of local or regional interest, while others, such as the enhanced greenhouse effect, concern all countries in the world. However, these global problems do have different effects in different locations. In the case of global climate change, for instance, Bangladesh and the low lying islands in the Pacific Ocean will be affected by flooding before many other countries feel any impact. Even on a European scale this distributional asymmetry is evident. Acidic deposition, for example, particularly affects forests on crystalline soils in Scandinavia and Central Europe, while forests on limestone are more resilient. So even reflections upon the importance of international and global environmental problems are influenced by location. Such variations will lead to different perspectives on the appropriate aims for local versus global society in the light of competing demands on limited resources.

In the following chapters this whole complex of problems is addressed from a variety of perspectives. The uniting issue is the focus upon interactions between environmental policies on the one hand, and the formulation of societal aims on the other. As outlined above, many questions arise in this debate. For example, can some variation in the use of policy tools, such as taxation and tradable pollution permits, be allowed between the regional and the national or international level? That is, to what extent can policy tools be implemented at different levels of the policy hierarchy to address different goals? How can we stimulate the participation of different groups in the process of the design and articulation of environmental policies? Is the local level more appropriate for this than the national or international? How can we deal with the interface between environmental policies and the semi-independent development of technologies? Does discussing and criticising modern technological developments on their environmental effects, before these effects are fully recognised by the general public, make sense in a modern democratic system? How can ecological economics take into account the relationship between other societal aims, such as income distribution and poverty abatement, and environmental problems?

The contributions in this book set out to discuss and offer answers to these questions. In the first part the focus is on local development and environmental policy. The second part concentrates on other policy goals, such as employment or economic growth and environmental policies, while the third part gives special attention to technological change. Finally, environmental policy tools and decision-making are given attention in the fourth part. Next we discuss the various contributions as they fall under these broad headings.

Local Development and Environmental Policy

The ways in which ideas of sustainable development can be hampered or facilitated by taking the regional context into account are discussed by Mack. Regions are affected strongly by global developments, in particular changing economic circumstances. For example, farmers in regions where extensive agriculture has been the normal practice are increasingly confronted with competition from farmers in areas with more favourable conditions for modern agriculture. Interest groups, such as these farmers, continually face trying to neutralise what they regard as negative external influences in order to maintain some stability in their lives. Mack argues that this can only be realised at a local or regional level because this is the level at which people can create and experience a cultural identity which allows them to realise the connections between ecological and economic systems. Thus, while sustainable development is a global issue, popularised by the World Commission on Environment and Development, achieving the changes it implies must be realised at a local or regional level. Local Agenda 21, as formulated at the Rio conference in 1992, provided the potential for local groups to promote sustainable development within the context of their own local circumstances. Rural identity, for example, is then seen as providing a 'sense of place' which enables interest groups to consider the challenges of sustainable development on a local basis.

Krarup's chapter addresses some similar issues concerning the role of local authorities in promoting sustainable development in a context of decentralisation. The extent to which decentralised decision making is required for efficient and effective environmental policy remains a significant problem. Indeed, the framework for most environmental policy is established by national or international institutions. However, in many cases, local authorities are better informed about local circumstances, which allows them the opportunity to realise serious improvements in the implementation of environmental policy. The case for decentralisation of environmental decision making can be based on variations in the ambient environment across regions, which imply different marginal damage functions and abatement costs. In addition, as some environmental functions are local, public goods decentralisation seems appropriate. However, local environmental responsibility can entail conflicts of jurisdiction. Based on a Danish survey, Krarup concludes that local decentralisation is likely to benefit the environment if municipalities have adequate resources and an advantage in terms of their knowledge of local circumstances. However, if other social problems are perceived to exist, such as unemployment, local authorities are inclined to give these priority over environmental problems and this will result in poor implementation and enforcement of environmental policies.

Corbett and Voisey address the possibility of using an institutional framework to realise sustainable development without frustrating the goal of local equity. They pay special attention to social vulnerability in managing sustainability at a local level. Vulnerability, as defined in light of the entitlement approach attributed to Sen, is differentiated from quantitative poverty. This means that the assets possessed by local actors, be they either poor or wealthy, play a significant role in determining their reaction to a sustainability-enhancing policy. These assets in turn depend upon the local institutional framework. This means that the analyst must incorporate the local institutional framework in order to understand the process of local or regional sustainability. Thus, the analyst must discover the extent to which current institutions can hinder sustainable development goals. In their approach to regional issues, Corbett and Voisey propose criteria for assessing the likelihood of conflicting interests, and use empirical examples including the European fishing industry.

Environmental Policy Versus Other Economic Policy Concerns

Hansen discusses the relationship between the aims of environmental policy in realising sustainable development and central issues from monetary theory. In macroeconomic theory, rule on the appropriate level of savings have traditionally attracted much attention. Thus, a minimum level of savings has been proposed as necessary in order to guarantee the future economic growth of any national economy. The discussion of the criterion for sustainable development has regenerated this debate. The core question is whether there are any savings rules which should be followed to guarantee or facilitate sustainability. This question has much in common with the debate on 'greening' national accounting procedures, because one of the central concerns is about which part of Gross National Product (GNP) can be valued as the sale of assets as opposed to the creation of sustainable income. The sale of assets has to be treated as capital consumption or depreciation. The depreciation part of GNP has, in turn, to be subtracted from the traditional measure of GNP to realise a 'green' type of GNP measure.

In this debate on the definition of assets, the environment is included as categories of natural capital, such as exhaustible resources, renewable resources and environmental resources. There is some danger in reducing the environment into these apparently objective and technical groupings. The ozone layer, atmospheric permeability for heat radiation, clean air, soil, and drinking water are all life and health support functions. Impairing these functions will affect future production and in the extreme failure to maintain the functions will prevent survival. This implies that the appropriate level of these functions has to be defined, as otherwise no calculations can be made regarding the environmental costs of the current production processes. However, no value-free definition of these functions is possible. While economists often make similar judgements about human life, the implication is that an ethical debate must take place.

Another problem is the extent to which reproducible capital can replace the natural resources which are used up in the current production processes. Hansen discusses which of the specific functions of natural resources used in the economy can be replaced by additional national accumulation of reproducible capital. He argues that sustainability criteria which are built on non-declining wealth cannot be measures for sustainable development. However, he believes they can be extended to such indicators, taking the specific goals of sustainable development into account by doing so.

Harribey explains that environmental problems cannot be discussed without paying full attention to distributional issues. On the basis of Georgescu-Roegen's work he argues that the limits to the growth potential of human society are clear. In fact, economists have long argued that there are only limited amounts of resources available to meet a variety of competing wants. Rawls' work on social justice is then used to argue for the need to clarify the relationship between the level of income and environmental limits. In order to give his approach a theoretical basis, Harribey proposes a simple mathematical model for the reduction of the scale of income while making an unemployment decrease possible and protecting ecological systems from the threat of excessively high economic growth. The model can function in two ways. First, for a given unemployment rate, it allows the determination of the scale of inequality reduction to be achieved in order to cut unemployment within a given period. Second, it can determine the time necessary to reduce inequality as a function of a decision about the socially desirable and sustainable income hierarchy. The model is then tested for the French economy. Harribey argues that unemployment elimination is a distributional problem rather than one due to the level of wealth. In the case of moderate growth, social and cultural changes can be triggered in ways that improve welfare. The resorption of unemployment is essentially achieved by the reduction of work time and the redistribution of income, while the preservation of nature and the environment is achieved by a moderate economic growth rate.

Kephalicos and Grimal intertwine the issues of innovation and technology with that of equity through job creation. They address the issue of employment and sustainable development, building on the notion of 'qualitative growth', defined by the decrease of natural capital used in the production function. This can only be attained by the mitigation of the strong complementarity between natural capital and other types of capital. Job creation can occur if firms adopt proactive strategies which try to cope with or even anticipate environmental regulations. The orientation towards qualitative growth means that firms upstream activities, aimed at designing cleaner production processes, must prevail over downstream activities, aimed at containing pollution damages by end-of-pipe pollution abatement. However, a comparison between France and Germany seems to indicate that end of pipe solutions predominate. Furthermore, the impression is given that the specialisation in downstream activities, and the corresponding employment pattern, could lead to an 'unemployment trap', as countries specialised in upstream activities, with the corresponding bias towards

Research and Development, export clean technologies to other countries and wipe out dirty ones with high abatement costs.

Technological Change and Environmental Policy

Cohen believes that the concept of ecological modernisation, as defined by German researchers in the field of environmental policy during the 1980s, is the central issue of the debate on the role of technological change. These researchers argue that currently polluting sectors, and basic industries in particular, are taking a more proactive position regarding environmental policies and have recognised that society no longer accepts an ongoing deterioration of nature and the environment. Therefore industry is regarded as trying to change production processes in such a way that production growth is accompanied by a decrease in the emissions of polluting substances and material use. In doing so, firms are also able to modernise production.

The concept is, in principle, based upon the assumption that strict government regulations can have a beneficial influence on the development and implementation of innovative processes. Along these lines, domestic producers will be able to retain a 'first-move' advantage in export markets for environmental technologies and 'green' products. This can only be realised if these industries introduce integrated pollution management accompanied by the use of anticipatory planning practices. Constructive relationships among industry, government, and the public are a precondition.

Cohen investigated the extent to which this type of ecological modernisation can be recognised. Elements of such approaches can be found in Germany, the Netherlands, and the USA. A general assumption, questioned here, is that societies which have a strong tradition of policy formulation involving all stakeholders have greater opportunities for ecological modernisation. Cohen finds that, in the Scandinavian countries, which have a high level of involvement of stakeholders, ecological modernisation is difficult to recognise. Cohen finally argues that further research in this field is necessary to develop more concrete criteria.

Ferrante deals with the relationship between technological change at the local level and an efficient policy on global warming. A model is used which assumes that knowledge spillovers between different production processes are incomplete and that firms face the same relative prices for inputs and the same technological opportunity set. Significant imperfections in the input markets or a substantial degree of technological heterogeneity may result in a quite different impact on the evolution of production possibilities. On the other hand, in a context of technological competition and evolutionary selection, localised technical change should emerge as a long-run outcome. The model is used to conclude that technological and institutional inertia are two constraints on pollution control. A status quo approach may lead to an underestimation of the real technological opportunities for mitigating global warming at low or even negative costs to society and overestimate the political feasibility of price-induced changes in energy production and consumption patterns. Although the analytical framework presented here fails to provide a ready-to-use operative tool for evaluating different pollution control strategies, the model shows that institutional as well economic constraints appear to play an interlinked role in the analysis of long-term strategies of pollution control. In future research these connections and relationships have to be given more attention.

The Societal Scope of Environmental Policy Tools and Decision making

Rienstra and Nijkamp discuss the environmental problems related to the transport sector. They argue that transport systems ignore the environmental problems resulting from the use of different modes of transportation. All individual users of a mode of transportation only take into account those effects that are relevant for the individual decision maker. Negative side-effects of these decisions are ignored as these effects are only recognised on a societal basis. As a result, free riding is a common practice. In addition, many individuals face a lack of adequate information on alternative possibilities. Only a procedural rationality is taken into account, which implies that habit formation is normal. This is particularly true for car drivers as this mode of transport is multi-functional. Car owners often exclude other modes of transportation as an option.

Public policy is then seen as aiming to neutralise these tendencies. Rienstra and Nijkamp explain this by using the principles of public choice theory. They argue that government policies in general, and thus also environmental policies, are often designed to win elections so as to attract the average voter. By doing so, the vested interests of certain groups are often protected by these policies. Furthermore, civil servants are likely to seek the maximum level for their budgets by releasing only part of the information on the costs and effects of the contemplated measures to the political powers. This will result in sub-optimal situations in which environmental targets are never realised.

The behaviour of pressure groups and stakeholders can be described by collective action theory which argues that environmental pressure groups and lobbies are often weak given their large size and diffuse objectives. Other interest groups are often better organised, have more support in traditional economic circles and so strong connections with parts of the state machinery in the economic sphere. These pressure groups include industrial lobbies, railway companies, associations of car drivers and the oil industry. Often the employment argument is used effectively against environmental controls and particularly so in periods and places of high unemployment.

The result of these different forces is that, although there is a general recognition of the necessity of reducing the environmental costs of transportation systems, there is little evidence that stakeholders agree on the ways and means to address transportation problems. The only possibility is to generate a level of information that would produce high levels of cognitive dissonance, the existence of appealing alternatives and, finally, a cultural change of attitudes.

Alio, Kendall and Laredo address the issue of local civil participation, especially in the process of Environmental Local Audits (ELAs) as management devices, for addressing local environmental issues. ELAs are only useful if carried out in the context of an ecologically and socially open-minded vision. This requires recognising two things. First, that initiatives for environmental change will fail without the support and engagement of the population. Second, that environmental reform requires changing economic models which are seen as the basis of many of the detected problems which society is facing. Thus, citizen participation and involvement is important, not only because it is a way to achieve political consensus, but also because it contributes to a change in ethical values and attitudes towards the environment. Moreover, they highlight the importance of the territorial dimension of sustainability and argue that using the geographic concept of territory enables the articulation of local issues as well as a global framework. ELA's are then seen as indicators of the local potentials.

Dijkstra and Nentjes discuss the influence of strong interest groups on the choice of instruments in environmental policy. In general, the economic theory of environmental externalities argues that market oriented instruments outperform direct regulation. However, in practice, direct regulation is still the most favoured instrument. A possible reason for the popularity of direct regulation is that powerful interest groups, such as capital owners and workers, expect to be worse off with market instruments than with direct regulation and have been successful in blocking the widespread use of market instruments.

Drawing on the work of Buchanan, the authors construct a new model which enables them to analyse the preference of interest groups over different policy tools given assumptions about those groups basic interests. Thus, workers are assumed to be motivated by employment prospects; they prefer direct regulation to market instruments. When long-run demand for output is elastic, direct regulation in the form of standards offers greater employment opportunities than a policy based upon pollution control 'bubbles'. With inelastic demand, standards might still be the preferred instruments if the workers' rate of time preference is high, otherwise pollution control 'bubbles' offer the highest employment level in the long run. Consumer and worker interests are identical. With elastic demand, lower prices and more employment go together. Consumers will always prefer direct regulation to market instruments. Grandfathering of permits is the most preferred instrument for shareholders, since it maximizes profits in the short run, and it is the only instrument that yields extra shareholder revenues in the long run. Pollution control 'bubbles' are the second preference of shareholders, while standards are given third place. Emission charges and an auction of tradeable permits rank lowest.

Thus the model explains the persistence of direct regulation in environmental policy. It shows that the market instrument of grandfathering of tradeable permits has one advocate, namely the shareholders. The acceptability of market instruments that yield government revenue (i.e., tradeable permits and emission charges) depends critically upon the possibility for redistributing the revenues to reduce the resistance from various interest groups. There is thus a strong incentive towards the adoption of an explicitly sub-optimal policy from the efficiency perspective.

Gouzee and Willems discuss the possible role of an energy tax in the European Union. They recall the basic statements of the Rio conference about unsustainable production and consumption patterns which have produced the major risks to the global environment. They are of the opinion that the rumours that the political process has failed to implement an energy tax have been largely exaggerated. The European Commission's carbon/energy tax proposal is the first coordinated environmental tax ever proposed to a group of Nations. The difficulties and obstacles which have arisen are therefore unsurprising. If Europe fails to overcome the problems, given its scientific, political and financial capacity, the chance for other less developed regions of the world implementing a such taxes seems doubtful. However, the adoption of the carbon tax and control of the enhanced greenhouse effect require major shifts in economic systems. The authors claim that such important changes will take, at least, one generation.

2 INDUSTRY PREFERENCES FOR INSTRUMENTS OF ENVIRONMENTAL POLICY

Bouwe Dijkstra and Andries Nentjes
Department of Economics, Faculty of Law, University of Groningen, Groningen, The Netherlands

Abstract

Although economists have shown that market instruments outperform direct regulation, environmental policy still consists mainly of direct regulation. One reason for this could be that powerful interest groups prefer direct regulation. Following Buchanan and Tullock (1975), Dewees (1983), and Spulber (1985, 1989), this paper studies restriction of emissions in a perfectly competitive industry. The instruments reviewed are: tradeable emission permits, (grandfathered and auctioned), emissions charges, firm bubbles (restriction of emissions per firm), and performance standards (restriction of emissions per unit of output). The paper offers a comprehensive review of the short run and long run effects of these instruments on, *inter alia*, profits, employment and product price. These are key variables which are of importance for shareholders, workers and consumers. We conclude that all three interest groups rank an auction of tradeable permits and emissions charges lowest. Shareholders rank grandfathering highest. Consumers rank standards highest. Workers rank standards or bubbles highest.

2.1 Introduction

Since Pigou's pioneering work in 1920 and in particular from the early seventies on (e.g. Baumol and Oates 1971), economists have argued that when it comes to dealing with environmental externalities, market-oriented instruments outperform direct regulation. With pollution charges or tradeable pollution permits emissions can be reduced at lower total control cost than is possible with command and control methods. But up to now the policy advice derived from economic theory has had hardly any political impact. Direct regulation is still the most used instrument..[1] And although the importance of market-oriented instruments is rising slightly, the policy stance in Europe is more oriented to bringing flexibility into regulation than to introducing market instruments.

How can this state of affairs be explained? Buchanan and Tullock (1975) suggested an answer more than twenty years ago. Their analysis is based on the view that the widespread introduction of market instruments is blocked by interest groups from industry, in particular those representing the interests of capital owners. The motive for their opposition is that, on the level of firm and industry, market instruments have a more negative impact on profits than direct regulation has. To support their case, Buchanan and Tullock (1975) construct a model of a competitive industry where emissions can be reduced only by curtailing production. Two instruments to reduce industry emissions to the political target level are compared: direct regulation in the form of non-tradeable output quotas, and regulation by way of a charge on output. Their impacts on profits in the short and in the long run are analyzed. The conclusion is that in the short run the representative firm's losses are larger with output charges than they are with output quotas; in the long run output quotas might even yield profits if the government were to restrict entry.[2] Thus firms will prefer and lobby for direct regulation in the form of output quotas and will resists output charges.

Buchanan and Tullock's (1975) article was a landmark, opening the political economy of environmental policy as an area of investigation. But the article has its weaknesses; in particular it is too restrictive in scope. Dewees (1983) seized the opportunity and elaborated Buchanan and Tullock's (1975) approach, simultaneously widening its scope. First he assumes abatement is possible: given the level of output, emissions can be reduced by making abatement costs. Second, he differentiates between shareholders and workers as interest groups. Shareholders are identified as interested in maximum profits and workers in keeping their jobs. This implies that Dewees (1983) not only investigates the effect of instruments on profits but also their employment impact. Third, besides direct regulation, in the form of a mandated emission to output ratio or performance standard, and emission charges, he discusses tradeable emission permits, distinguishing between auction of permits and handing them out free to established firms (grandfathering). In our view these changes can be seen as scientific progress compared with Buchanan and Tullock (1975). As a step backward we consider Dewees' (1983) decision to take not the restriction of

---

[1] See OECD (1994) for an overview of market instruments in practice.

[2] In fact, with Buchanan and Tullock's (1975) treatment of emissions charges, one cannot be sure that industry will always prefer direct regulation (Dijkstra 1994).

total industrial emissions to a given level as the goal of environmental policy but instead a maximum emissions to output ratio. The consequence is that total industrial emission levels differ, as industrial output differs across instruments. The upshot of Dewees' (1983) model is that shareholders will prefer grandfathering of tradeable permits to standards, and standards to emissions charges or an auction of tradeable permits. It should be noted that the introduction of tradeable permits which are grandfathered places a market instrument above direct regulation. It can be shown that this result also holds in Buchanan and Tullock's (1975) model, if it is extended.[3] Basically Dewees (1983) and Buchanan and Tullock (1975) do not differ in ranking preferences of shareholders and of firm management in so far as they are maximizing shareholders' value. The workers' interest in instruments differs from that of shareholders. Employment will be hurt less if environmental policy is implemented with direct regulation in the form of performance standards than it is if market instruments, either emission charges or tradeable emission permits, are applied. Therefore workers will prefer performance standards to market instruments.

Although Dewees' (1983) analysis meant progress, it leaves a number of questions unanswered. A first question is whether his ranking still holds if total emissions by industry are not allowed to vary across instruments, and controlling industrial emissions to a specific level is taken as the government goal. Besides that, the set of instruments should be extended. Environmental agencies have experienced that setting performance standards is a rather rigid form of direct regulation and in some cases they have made it more flexible by direct regulation in the form of non-tradeable emission quotas per firm or firm bubbles. Firm bubbles can also be interpreted as a representation of an instrument widely used in the Netherlands: the covenant. A covenant is a (voluntary) contract between the government and a (group of) firm(s) to reduce emissions to an agreed maximum. A third point of criticism of Dewees (1983) is that he makes very specific assumption about the technology of the firm. He assumes abatement is only possible with end-of-pipe technology, and abatement does not decrease the optimal scale of a firm. Again the question is how far his ranking order follows from these specific assumptions.

An important contribution to answering these questions has been made by Spulber (1985, 1989).[4] He does not directly investigate interest group

---

[3] See Dijkstra (1994), who also reviews the discussion on this subject following Buchanan and Tullock's (1975) article (Coelho 1976, Yohe 1976, Buchanan and Tullock 1976).

[4] Spulber (1985, 1989) offers the most thorough treatment of the effects of environmental policy instruments. Dijkstra (1998b) also reviews the rest of the literature on this subject, including

preferences, but his analysis is relevant to the debate on the political choice of instruments in environmental policy. Spulber (1985, 1989) analyzes the long-run effects on product price, number of firms, emission per firm, production per firm, and welfare, of a broad range of instruments, including tradeable emission permits, emissions charges, effluent standards or quotas (which we shall call firm bubbles), output taxes and entry fees, subsidies, and technological standards. Unlike Dewees he takes the quantity of industry emissions as given. Besides that he places fewer restrictions on the production and abatement technology.

In this article we build on the analysis of Dewees (1983) and Spulber (1985, 1989) in reconsidering the question of how different interest groups rank the various instruments of environmental policy. Thus the question is: which instrument would be the interest group's first choice, what is second, and so on. Our aim is to derive these rankings in an industry model that is as general as possible. The major new elements in the analysis are:

- the assumptions about production and abatement technology are more general. The restrictions on the cost function are not imposed to facilitate the subsequent analysis, but they are derived from considerations like the second order condition for profit maximization (Dijkstra 1998b);
- emissions are kept constant at the emission target level for industry;
- the ranking order for interest groups is based on short and long run impacts, unlike Spulber, who considers only long run effects; and
- three interest groups are considered: in addition to shareholders and workers we distinguish consumers.

Summing up, the aim of our analysis is to be broader (short and long run, more interest groups) as well as deeper (more general cost function, constant industry emissions) than earlier research. The rest of this paper is organized as follows. The assumptions of our analysis are set out in Section 2.2. Sections 2.3 and 2.4 contain a description of the short run and the long run industry response to environmental policy instruments. Whatever the instrument, it is assumed that once the short run, respectively long run equilibrium of industry is realized, the target level of total emissions is also realized. The results of the analysis are summarized in Table 2.1. Section 2.5 discusses the welfare impacts of the instruments. Section 2.6 discusses the preference ordering of instruments by

---

Hochman and Zilberman (1978), Harford and Karp (1983), Besanko (1987), Helfand (1991), and Helfand (1993).

interest groups: the consumers, the shareholders and the workers. The conclusions follow in Section 2.7.

2.2 Assumptions

We will examine the effects of the following instruments: first, direct regulation in the form of: (a) performance standards (also called standards), a maximum ratio of emissions to output and (b) firm bubbles (also called bubbles), a ceiling on the total emissions of a firm. Second, market instruments in the form of: (a) emissions charges, a charge per unit of emission and (b) tradeable emission permits, permits for a unit of emission. These permits can be either auctioned or grandfathered. Emissions charges and the auction of tradeable permits are also called financial instruments. We look at the effects of the instruments upon an industry that is characterized by perfect competition between firms with identical cost functions. Before environmental policy is introduced, the industry is in long run equilibrium without any government intervention.

By choosing perfect competition with identical firms as our market model, we eliminate two differences between market instruments and direct regulation, which actually do play an important role in the discussions on policy instruments. First, the static efficiency advantage of market instruments (i.e. that emissions are distributed in a cost-minimizing way across firms) is eliminated, because all firms are equal, not only in production technology but also in their abatement cost functions. Second, the dynamic efficiency advantage (i.e. that market instruments promote innovation in abatement technology) is eliminated, because we abstract from technological progress.

In the first step (Sections 2.3 and 2.4), the analysis concentrates on the effects of environmental policy on prices (relevant for consumers), number of firms and production and emissions per firm. From these we can derive the effects on profits and employment, that are of interest to shareholders and workers respectively.

We distinguish a short run and a long run response to environmental regulation. In the short run, the number of firms is given, but production per firm is variable. In the long run, firms can enter and leave the industry at no cost, so that economic profit is zero in equilibrium.

The cost function we employ has the general form $C(q,E)$, where q is production and E is emissions per firm. Although the short run cost function does differ from the long run cost function, we will use the same general notation for both. Dijkstra (1998b) derives the restrictions to be placed on the cost function. These restrictions stem from a number of considerations, including second order conditions for profit maximization and monotonicity: the condition that industrial emissions are declining in the strictness of the instrument. The following restrictions, derived in Dijkstra (1998b), are relevant for the analysis in the next sections:

$$C_q > 0$$
$$C_E \leq 0$$
$$C_{qq} \geq 0\,,\ C_{Eq} = C_{qE} \leq 0;\ qC_{qq} + EC_{qE} \geq 0$$

2.3 The short run

Since the number of firms n is equal for all instruments in the short run, emissions per firm E must be equal for all instruments to keep industry emissions equal. When the government applies firm bubbles, it simply restricts emissions per firm to E. When the instrument is tradeable permits, the government issues permits allowing industry emissions of nE. In case of emissions charges, the government has to set the charge at the level at which the firm chooses to emit E. When standards are used, the government has to set the emissions to output ratio at the level at which the firm chooses to emit E.

2.3.1 Bubbles and market instruments

In the case of firm bubbles, the firm maximizes profits with a restriction of E' on its emissions:

$$\max Pq - C(q,E) - \lambda(E-E').$$

When the firm has to pay an emission charge of t, it maximizes operating profit minus the charge bill:

$$\max Pq - C(q,E) - tE.$$

When the instrument is tradeable permits, the firm maximizes operating profit minus expenditures on permits (where E_0 is the amount of permits grandfathered to the firm, and t' the permit price):[5]

---

[5] To be precise, the amount $t'E_0$, which is the value of the grandfathered permits, is an opportunity cost to the firm, but a revenue to the shareholders. Since this amount is a revenue to shareholders,

$$\max Pq - C(q,E) - t'(E-E_0).$$

We see that, with firm bubbles as well as with the market instruments of emissions charges and tradeable permits, the first order conditions are the same:

$$P - C_q(q,E) = 0\ (1^b),\ (1^m);\ \lambda = t = t' = -C_E(q,E).$$

The product price equals marginal cost of producing output, keeping emissions constant. The marginal cost of emission reduction equals the price of emission, either in the form of an emissions charge, a permit price, or the shadow price with a bubble. Thus, market instruments and firm bubbles yield exactly the same production per firm and product price. The reason for this is easily seen when comparing firm bubbles to the market instrument of grandfathering (with $E = E_0$). Both instruments give a firm the right to pollute E. The only difference is that with grandfathering, this right is tradeable, while with firm bubbles it is not. But since all firms are equal and the number of firms is given, there will be no trade in the short run and firm bubbles are equivalent to grandfathering.

2.3.2 Standards

The instrument of performance standards resulst in a different first order condition than (1^b) and (1^m). The firm maximizes profit under the restriction that the emissions to output ratio does not exceed the allowed maximum e:

$$\max pq - C(q,E) - \lambda\ (E-eq).$$

The first order condition is: $P - C_q(q,E) - eC_E(q,E) = 0$, ($1^s$). Because $C_E(q,E) < 0$, we see that with standards: $P - C_q(q,E) < 0$. Since $C_{qq}(q,E) \geq 0$, q must be higher and p must be lower under standards than under the other instruments (remember that E must be equal). The reason for this result is that given q and E, the marginal cost of producing an extra q is lower with standards. When firms increase q, they are allowed to increase their emissions as well. When firms increase q with firm bubbles, they have to incur the additional cost of keeping E constant.

2.4 The long run

we shall add it to the economic profit, and call the sum 'short run profit', which is somewhat imprecise.

2.4.1 The long run equilibrium

In the long run, the number of firms is variable. The equilibrium number of firms will be such that the remaining firms make zero economic profit. In the long run, there are four equations for every instrument, that determine the equilibrium value of the product price P, production per firm q, emissions per firm E, and the number of firms n:

$$P = C_q \qquad (1^m),(1^b) \qquad P = C_q + (E/q)C_E \qquad (2.1^s)$$

$$Pq = C(q,E) - EC_E \qquad (2^m) \qquad Pq = C(q,E) \qquad (2.2^b),(2.2^s)$$

$$nE = L \qquad (2.3)$$

$$P = P(nq) \qquad \text{with } P'(nq) < 0 \qquad (2.4)$$

The superscript 'm' stands for market instruments, the 's' for performance standards, and the 'b' for firm bubbles. Equation (2.1) is the profit maximization condition, that has already been derived in the analysis of the short run. Equation (2.2) is the zero (economic) profit condition. Under direct regulation, firms should cover their operating cost C(q,E). When market instruments are used, there is an extra cost component $-EC_E$: the cost of the right to emit E. When the market instrument is emissions charges or an auction of tradeable permits, a firm must pay the government for this right. But when the market instrument is grandfathering, the firm must cover $-EC_E$ as well. The firm should include the opportunity cost of grandfathered permits in its calculation of cost. From the point of view of the firm, costs with grandfathered permits are not lower than with a permit auction. If a firm would only cover its operating costs, it had better sell its permits and leave the industry.

Condition (2.3) states that industry emissions for all instruments equal the government target of L. Condition (2.4) is the inverse market demand curve for the product.

2.4.2 Comparing standards to bubbles

In Section 2.5 we will show that standards are the welfare maximizing instrument of direct regulation in the long run. When direct regulation is applied, welfare in the long run only consists of consumers' surplus. The lower the product price, the higher consumers' surplus. Thus, the finding that standards yield a higher welfare level than bubbles implies:

$$P^s < P^b.$$

The intuition behind this is as follows. Consider what would happen if, starting from long-run equilibrium, the system of standards were changed into firm bubbles, keeping emissions per firm constant. This is illustrated in Figure 2.1. D/n is market demand, divided by the number of firms. AC and MC represent average and marginal cost for standards (s) and bubbles (b). Long-run equilibrium with standards occurs in point A. With bubbles, average cost is lower for $q < q^s$ and higher for $q > q^s$. The reason is that when firms decrease q with bubbles, they are allowed to keep E at the same level. With standards, E must decrease with q. But when firms increase q with bubbles, they must keep E constant. With standards, E is allowed to increase with q. For the same reason, marginal cost is higher for bubbles than for standards. The switch from standards to bubbles yields a short run equilibrium at C. Firms cut back their production. Product price rises and firms make a profit of BCFG. This profit will attract new firms. Allowed emissions per firm will be adjusted downward to accommodate the new firms. Profits per firm will decline with allowed emissions, until they are back to zero again. In the new long run equilibrium with firm bubbles, there will be more firms than with standards:

$$n^b > n^s$$

$$E^b < E^s$$

As new firms enter, product price will decline again. But it will not decline below the initial P^s, the long run price level with standards.[6] From $P^b > P^s$ it follows that $n^b q^b < n^s q^s$. Combining this with $n^b > n^s$:

$$q^s > q^b.$$

[6] Suppose that it would. This would mean that with firm bubbles, firms would still be making a profit at P^s, with n^b exceeding n^s and thus q^b below q^s, but with E^b/q^b equal to E^s/q^s. This is impossible, because given E/q, average cost is at a minimum, equal to P^s, at q^s. Thus, at $q < q^s$, there is a loss.

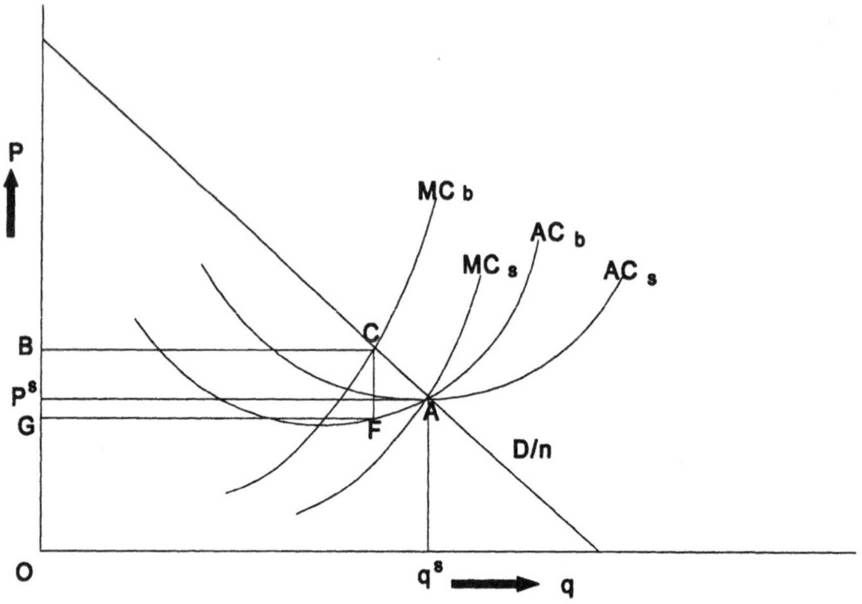

Figure 2.1 The switch from standards to bubbles

2.4.3 Comparing bubbles to market instruments

From economic analysis, it follows that:

$$n^m < n^b$$

and

$$E^m > E^b.$$

To see this, note what would happen if charges were replaced by bubbles. Firms would receive the non-tradeable right to emit as much as they did under charges. They would keep producing the same quantity of output, since the first order condition (1) is the same for emissions charges and firm bubbles. Then they would make a profit, because they would no longer have to pay the charge bill. The profit will induce entry and consequently the number of firms in long run equilibrium will be higher under standards and emission per firm will have to be lower than they are with market instruments. Furthermore, we can show that:

$$q^m > q^b.$$

This follows from the fact that $q^m \leq q^b$ is impossible. If $q^m \leq q^b$, then $P^m > P^b$, from $n^m < n^b$ and (4). According to (1), $P^m > P^b$ implies $C_q{}^m > C_q{}^b$. But $q^m \leq q^b$ and $E^m > E^b$ imply $C_q{}^m \leq C_q{}^b$ since $C_{qq} \geq 0$ and $C_{qE} \leq 0$. Finally,$P^m \geq P^b$. To see that $P^m < P^b$ is impossible, we look at:

$$\frac{dC_q}{dq} = C_{qq} + \frac{E}{q} C_{qE} \frac{dE/E}{dq/q} \tag{2.5}$$

For $P^m < P^b$, the left hand side of the expression should be nonpositive, in view of $q^m > q^b$ and (1). But $P^m < P^b$ also means that $E^m/q^m < E^b/q^b$, because of (2.3) and (2.4). Thus, going from q^b and E^b to q^m and E^m:

$$\frac{dE/E}{dq/q} < 1.$$

This inequality, together with those from Dijkstra (1998b), implies that the right hand side of (2.5) will be positive. Thus, we have generated an impossibility.

2.4.3 Comparing standards to market instruments

From $P^m \geq P^b$ and $P^b > P^s$ it follows immediately that:

$$P^m > P^s$$

and

$$E^m/q^m > E^s/q^s.$$

From this last inequality we can derive that:

$$E^m > E^s$$

and

$$n^m < n^s.$$

This is because with standards as well as with market instruments, we can derive from (1) and (2) that in the long run equilibrium, the cost function satisfies the homogeneity condition:

$$C(q,E) = EC_E + qC_q. \tag{2.6}$$

This means that, when increasing the strictness of environmental policy from zero to prohibitive, (q,E) follows the same path with both instruments, only with a different speed. Since monotonicity implies that E/q declines in the strictness of environmental policy, and $E^m/q^m > E^s/q^s$, (q^s,E^s) is 'ahead of' (q^m,E^m) for any given level of L. Dijkstra (1998b) shows that $dE/dL \geq 0$ for all instruments. Thus,

$$E^m > E^s.$$

We need a mild extra assumption

$$C_{qE} + (E/q)C_{EE} < 0$$

to ensure that $dq / dL \geq 0$ for standards and market instruments. Since, as we have seen, (q^s,E^s) is 'ahead of' (q^m,E^m):

$$q^m > q^s.$$

Table 2.1. Effects of instruments

	short run	**long run**
product price	$P^s < P^{oi}$	$P^m > P^b > P^s$
number of firms	$n^s = n^{oi}$	$n^m < n^s < n^b$
emission per firm	$E^s = E^{oi}$	$E^m > E^s > E^b$
production per firm	$q^s > q^{oi}$	$q^m > q^s > q^b$
s = standards oi = other instruments	b = firm bubbles	m = market instruments

5. Welfare

Total welfare consists of consumer surplus, economic profits, and government revenue. We want to compare the welfare effects of the instruments under the restriction that all instruments yield the same total industry emissions L. In order to find out which instrument maximizes welfare W, we maximize the following expression:

$$W = \int_0^{nq} P(y)dy - nC(q,E) - \lambda(nE-L) \tag{2.8}$$

The first term on the right hand side of (2.8) is the area below the inverted demand curve P(nq). To obtain total welfare, we have to deduct industry cost (which is the second term) from this first term, and we add the constraint that industrial emissions do not exceed L. Bearing in mind that the short run cost function differs from the long run cost function, expression (2.8) applies to the short run and the long run. The short run conditions for welfare maximizing differ from the long run conditions, because in the short run, n is fixed, while in the long run, n is variable. Thus, in the short run version of (2.8), E is fixed, since n and L are fixed. The only variable is q, which determines P. Maximizing (2.8) with respect to q:

$$nP(nq) - nC_q(q,E) = 0.$$

This condition is identical to (1^b) and (1^m): the first order conditions for short run profit maximization with market instruments and bubbles. It differs from (1^s), the condition for profit maximization with standards. Thus, market instruments and bubbles maximize welfare in the short run, and standards do not. In the long run, there is no economic profit, so that total welfare only consists of consumer surplus, and government revenue. For long run welfare, we maximize (2.8) with respect to n,q and E. The first order conditions are:

$$nP(nq) - nC_q(q,E) = 0$$

$$qP(nq) - C(q,E) - \lambda E = 0$$

$$nC_E(q,E) - \lambda n = 0$$

$$nE - L = 0$$

We see that the welfare maximum satisfies (1^m) and (2^m), the profit maximization and zero profit condition for market instruments. Thus, market instruments maximize welfare in the long run. This is a well-known result (Spulber 1985). Direct regulation cannot maximize long run welfare, because the zero profit condition deviates from (2^m). But one can also compare the welfare impacts of different instruments of direct regulation. To see which form of direct regulation maximizes long run welfare (which now only consists of consumer surplus), we maximize welfare under the additional restriction (2^b),(2^s), defining zero profit under direct regulation:

$$\max \int_{0}^{nq} P(y)dy - nC(q,E) - \lambda(nE-L) - \mu[qP(nq) - C(q,E)].$$

It can be shown (Dijkstra 1998b) that the solution to this maximization problem is (1^s), the first order condition for profit maximization with standards. Thus, standards are the welfare maximizing instrument of direct regulation in the long run. We have the counterintuitive result that in the short run, bubbles yield higher welfare than standards, whereas in the long run, standards yield higher welfare than bubbles.

The result of the analysis is summarized in Table 2.2. Market instruments and bubbles are welfare-maximizing in the short run. In the long run, only market instruments are welfare-maximizing, and standards result in higher welfare than bubbles. Thus, it is clear that market instruments rank highest on overall welfare. It is unclear how bubbles and standards compare on overall welfare.

Table 2.2. Welfare ranking of instruments

Short run	Long run	Final rank
1,2.market instruments, bubbles	1.market instruments	1.market instruments
	2.standards	2/3.bubbles or standards
3.standards	3.bubbles	

2.6 Interest groups

2.6.1 Consumers

Although consumers may have a small per capita stake, they are likely to voice their preferences if they are already united in an interest group. Consumers of a specific product are interested in low product prices; for example car drivers prefer low fuel prices. Therefore we assume the consumers will prefer the instrument that yields the lowest product price. The lower product price, the higher consumer surplus will be. The effect of instruments on product price is shown in Table 2.1. The ranking of the instruments by consumers that can be derived from this is given in Table 2.3. In the short run, standards are the consumers' favorite instrument. The consumers are indifferent between all other instruments. In the long run, standards are again ranked highest. Bubbles rank second and market instruments last. Thus, the consumers' overall preference ranking, shown in Table 2.3, is the same as the ranking on long run effects.

Table 2.3 Consumers' preferences for instruments

Short run	Long run	Final rank
1.standards	1.standards	1.standards
2,3.firm bubbles, market instruments	2.firm bubbles	2.firm bubbles
	3.market instr	3.market instruments

2.6.2 Shareholders

The shareholders receive a normal rate of return on the capital they have invested, plus the economic profit or loss. Let us first look at short run profits. When the instrument applied is a permit auction and emissions charges, and also when it is standards, profits can be written as:[7]

$$\pi = qC_q + EC_E - C(q,E).$$

Differentiating this partially to q yields:

$$\delta\pi/\delta q = qC_{qq} + EC_{Eq} > 0.$$

The inequality follows from (5a). We see that given E, profits are higher, the higher q. Thus, profits will always be higher with standards than with a permit auction and emissions charges, because given E, q is higher in case of standards.

It is clear that firm bubbles and grandfathering will always result in higher short run profits than do emissions charges and an auction of tradeable permits. All of these instruments result in the same level of production. The only difference is that in case of financial instruments, fimrs have to pay the government for the residual emissions, whereas they do not have to pay when grandfathering of permits or firm bubbles are applied. The amount that goes to the government with financial instruments, accrues to the shareholders with grandfathering and bubbles.

To compare profits under firm bubbles and grandfathering with profits under standards, we use the result from Section 2.5 that firm bubbles, and market instruments are the welfare maximizing instrument in the short run. But we have also shown that consumers' surplus is higher with standards (subsection 2.6.1). Total welfare being lower and consumer surplus higher, operating profits must be lower when standards are applied than in the case of bubbles and grandfathering.

We conclude that short run profits are highest with grandfathering and bubbles, lower with standards, and lowest with the financial instruments. This ranking is shown in Table 2.4.

Table 2.4. Shareholders' preferences for instruments

[7] In the case of a permit auction and emission charges we have, substituting (1^m): $\pi = Pq - C(q,E) + EC_E = qC_q + EC_E - C(q,E)$. When standards are applied, we have, substituting (1^s): $\pi = Pq - C(q,E) = qC_q + EC_E - C(q,E)$.

Short run	Long run	Final rank
1,2.grandfathered permits,	1.grandfathered permits	1.grandfathered permits
firm bubbles	2-5.firm bubbles, standards, permit auction, charges	2.firm bubbles
3.standards		3.standards
4,5.permit auction, charges		4,5.permit auction, charges

We have assumed that there were no entry and exit costs, so that shareholders are indifferent between their firm's leaving or staying in the industry, and economic profit is zero in the long run equilibrium. However, there is a long run gain for shareholders with one instrument, namely grandfathering of tradeable permits. Shareholders of a firm that leaves the industry will receive the revenues from the sale of the firm's permits. Shareholders of a firm that stays in the industry will demand that the firm make an operating profit on the grandfathered permits, equal to the permit price. This operating profit counts as an economic (opportunity) cost to the firm, but as a revenue to its shareholders.

Thus, with respect to long run gains, the shareholders prefer grandfathering, and are indifferent between all other instruments, as shown in Table 2.4. Due to this long run gain, grandfathering rises to the nr. 1 position for shareholders, whereas the rest of the ranking is based on short run profits. This final ranking is also shown in Table 2.4.

2.6.3 Workers

We will identify workers' preferences with industrial employment. As Dewees (1983) already pointed out, workers are adversely affected when industrial employment declines. The key assumption is that when workers are laid off, they cannot start working somewhere else at not cost. They may incur costs of searching for and adjusting to a new job or being unemployed. They may have sector-specific skills that will not be rewarded in a different industry. They may lose seniority positions or pension rights. The traditional welfare measure (2.8) does not take these costs into account. There it is assumed that workers can change employment without cost, in which case they would of course be indifferent between all instruments.

Whether incumbent workers also benefit from an increase in industrial employment is less clear. Dewees (1983) assumes that they do not benefit. We assume here that they do.

Let us now compare the employment effects of the instruments. It is easily seen that short run employment is highest with standards and equal for all other instruments. This is because E is the same for all instruments, while q is higher with standards, and equal for all other instruments. Table 2.5 shows the ranking of the instruments on short run employment.

To compare long run employment, we assume, following Dewees (1983), that the long run cost function is homothetic, that is, labour costs are a fixed percentage of operating costs. Industry employment is always lowest with market instruments. This is easily seen from Figure 2.2 which illustrates long run market equilibrium under different instruments. Under market instruments, product price is highest (as we have shown in Section 2.4), and industry output $Q = nq$ is lowest. Total cost consists of operating cost $C(q,E)$ (represented by the area OKJH), plus charges or permit cost (area KNBJ). It has been demonstrated in Section 2.5 that market instruments maximize welfare. Therefore the area ABJK must be larger than the consumers' surplus under standards (AEL). That is to say, consumers' surplus plus charge revenues with market instruments must exceed consumers' surplus with direct regulation. Since consumers' surplus is lowest with market instruments, average operating cost must also be lowest with market instruments. In other words, market instruments maximize welfare mainly because they are associated with low cost of production. However, from the point of view of employment, this is their weakness. Total production and average operating cost being lowest, total industry cost must be lowest under market instruments. In Figure 2.2, the area OKJH will always be smaller than the areas OLEF and OMDG, because K will be below L and M, and H will be to the left of G and F. Since employment is assumed to be a fixed percentage of total operating cost, employment will be lowest under market instruments.

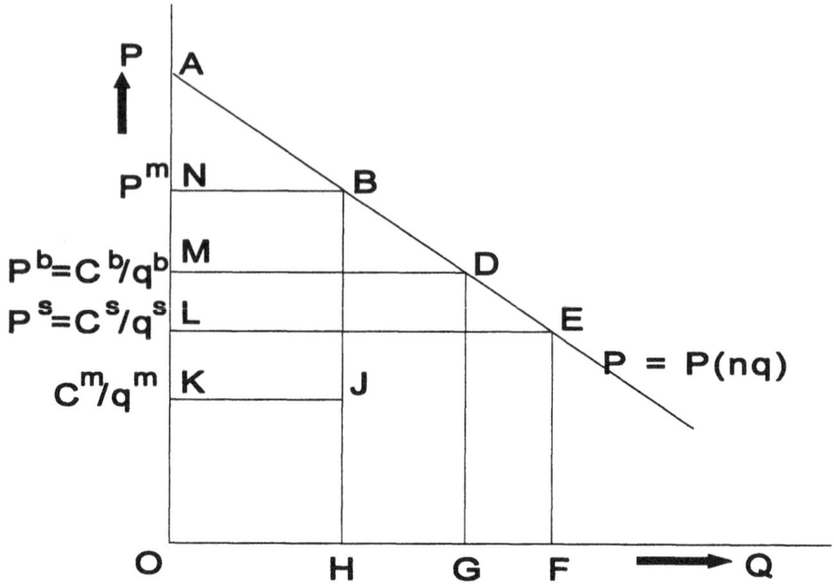

Figure 2.2 Employment in long run equilibrium

The difference between employment with standards and firm bubbles depends on the elasticity of demand. Remember that with bubbles, product price is higher. With inelastic demand, $|\varepsilon| < 1$, the amount of money that consumers spend on the product rises with the product price. In that case, bubbles will yield higher industry employment. With elastic demand, $|\varepsilon| > 1$, standards will yield higher industry employment. Table 2.5 displays the rankings on long run employment for elastic and inelastic demand.

Table 2.5 Workers' preferences for instruments

Short run	Long run		Final rank	
	$\lvert\varepsilon\rvert > 1$	$\lvert\varepsilon\rvert < 1$	$\lvert\varepsilon\rvert > 1$	$\lvert\varepsilon\rvert < 1$
1.standards	1.standards	1. bubbles	1. standards	1 or 2. standards or bubbles
2. bubbles market instruments	2. bubbles	2. standards	2. bubbles	
	3.market instruments	3. market instruments	3. market instruments	3. market instruments

We can conclude that workers, taking into account employment impacts in both the short and the long run, rank market instruments lowest. With elastic long run demand, they prefer standards to bubbles. With inelastic demand, it depends on the workers' trade off of short run and long run effects. This final ranking is alsc displayed in Table 2.5.

2.6 Conclusion

We have studied how various interest groups assess the instruments of environmental policy, the ranking of shareholders being based on the instruments's impact on profits, of workers on employment impact and for consumers on the effect on prices. The market instruments tradeable emission permits, with grandfathering and auction, and emissions charges were taken into consideration, as well as direct regulation in the form of firm bubbles and performance standards.

As for shareholders, grandfathering of permits is the most preferred instrument, since it maximizes profits in the short run, and is the only instrument that yields extra shareholder revenues in the long run. Firm bubbles come next. On the basis of short run profits they are preferred to standards, that come in third. Emissions charges and an auction of tradeable permits rank lowest. This concurs broadly with Buchanan and Tullock's (1975) and Dewees' (1983) rankings.

For workers, whose preferences are based on employment impact, direct regulation is always preferable to market instruments. When long run demand for output is elastic, direct regulation in the form of standards offers more employment than bubbles. With inelastic demand, standards might still be the preferred instrument if the workers' rate of time preference is high; if not it is firm bubbles, since they offer highest employment in the long run.

Table 2.6 Final ranking

<table>
<tr><th>Shareholders</th><th colspan="2">Workers</th><th>Consumers</th><th>Welfare</th></tr>
<tr><td rowspan="2">1.grandfathered permits
2. bubbles
3. standards
4,5. permit auction and charges</td><td>$|\varepsilon| < 1$
1 or 2. standards or bubbles</td><td>$|\varepsilon| < 1$
1.standards
2. bubbles</td><td rowspan="2">1. standards
2. bubbles
3,4,5. permits: grandfathered and auctioned, charges</td><td rowspan="2">1,2,3. permits: grandfathered and auctioned, charges
4 or 5. bubbles or standards</td></tr>
<tr><td colspan="2">3,4,5. permits: grandfathered and auctioned, charges</td></tr>
</table>

The impact of environmental policy instruments on product price is of interest to consumers. With elastic demand, lower prices and more employment go together. In that case consumer and worker interests are identical. More generally: workers and consumers will always prefer direct regulation to market instruments.

The rankings of the three groups are summarized in Table 2.6. It shows that when it comes to the political choice of instruments of environmental policy and if in such a case all options for instruments were on the agenda, shareholders and firm management as their agent would lobby for grandfathering of tradeable emission permits; but worker and consumer organisations would be against it and plead for direct regulation. When we confront this analysis with reality, we see that up to now the debate on instruments in Europe has mainly concentrated on the choice between direct regulation and charges. In the end regulatory charges have been introduced only in a very few cases. This could be interpreted as support for the conclusion that emission charges (and permit auction) are the instruments that are ranked lowest by all three interest groups.

Table 2.6 also reveals that bubbles are in the second place, or higher, for every interest group. This suggests that if interest groups try to come to a consensus, bubbles could be a compromise as a perhaps second best but politically feasible instrument. This is particularly relevant for the Netherlands. The impacts of the bubble on prices, profits and employment are represtentative of the effects of covenants or voluntary agreements. Nowadays the covenant is widely applied in the Netherlands; it might even be more important than standards. Table 2.6 indeed predicts that in a climate of consensus seeking, the bubble can expect broad support.

Although our analysis can explain a number of political choices that have actually been made, we should consider this model only as a building block of a more comprehensive analysis. For example, one has to qualify our result that the interest groups rank the financial instruments (emissions charges and an auction of tradeable permits) lowest of all. There are two factors that depress the interest groups' valuation of the financial instruments in our setup. First, we have assumed that market instruments do not have a static or dynamic efficiency advantage over direct regulation. This assumption had to be made to be able to get out mathematical results but one should not forget that the economist's argument for market instruments is based on their efficiency. Empirical studies support this view (Klaassen 1996). Second, we have assumed that the revenue of the financial instruments is used for general purposes from which the industry does not perceptibly benefit. When we take into account the possibility that industry can acquire part of the revenue, the acceptability of financial instruments will rise (Dijkstra, 1998a). Therefore it is quite possible that in specific cases, where efficiency gains and revenue distributions are taken into account, the financial instruments end up higher in the preference ranking of one or more interest groups.[8]

As a next warning it should be mentioned that our analysis of interest groups summarized in Table 2.6 is not complete. Other interest groups than the ones analyzed here will also be active. These are the environmentalist movement and the environmental bureaucracy. Their preferences and rankings, discussed in Nentjes and Dijkstra (1994) and Dijkstra (1998b), will also affect the actual political choice of the instrument of environmental policy.

In the introduction of this paper it was mentioned that the welfare economists' advice is to apply market instruments. We underpinned this position with conclusions derived from the model of a perfectly competitive industry and summarized the result in Table 2.2.The economist's position can be compared to the positions defended by interest groups. We see that the economists face a united front of workers and consumers. If it were not for the possibility of grandfathered permits, shareholders would also join the front for direct regulation in environmental policy. As we have mentioned, it is not impossible that they will accept the compromise of a bubble or covenant. Simple and incomplete though the model may be, it can explain the persistence of direct regulation, despite the economist's lament.

References

Baumol WJ, Oates WE. The use of standards and prices for protection of the environment. Swedish Journal of Economics 1971; 73: 42-54.

Besanko D. Performance versus design standards in the regulation of pollution. Journal of Public Economics 1987; 34: 19-44.

Buchanan JM, Tullock G. Polluters' profits and political response: direct controls versus taxes. American Economic Review 1975; 65: 139-47.

Buchanan JM, Tullock G. Polluters' profits and political response: direct controls versus taxes: reply. American Economic Review 1976; 66: 983-5.

Coelho PRP. Comment on polluters' profits and political response. American Economic Review 1976; 66: 976-8.

Dewees DN. Instrument choice in environmental policy. Economic Inquiry 1983; 21 : 53-71.

Dijkstra BR. Polluters' profits: direct controls versus taxes versus tradeable permits. Buchanan & Tullock revisited. Research Memorandum 12, Department of Economics and Public Finance, Faculty of Law, University of Groningen, 1994.

[8] See Dijkstra and Nentjes (1998) for a discussion of these issues in the context of a post-Keynesian model of the firm.

Dijkstra BR. A two-stage rent-seeking contest for instrument choice and revenue division, applied to environmental policy. Forthcoming in European Journal of Political Economy, 1998a.

Dijkstra BR. *The political economy of instrument choice in environmental policy.* PhD dissertation, University of Groningen, 1998b.

Harford JD, Karp G. The effects and efficiencies of different pollution standards. Eastern Economic Journal 1983; 9: 79-89.

Helfand G. Standards versus standards: the effects of different pollution restrictions. American Economic Review 1991; 81: 622-34.

Helfand G. The relative efficiency of different standards when firms vary. Natural Resource Modelling 1993; 7: 203-17.

Hochman E, Zilberman D. Examination of environmental policies using production and pollution microparameter distributions. Econometrica 1978; 46: 739-60.

Klaassen G. *Acid rain and environmental degradation: the economics of emission trading*. Cheltenham: Edward Elgar, 1996.

Nentjes A, Dijkstra B. The political economy of instrument choice in environmental policy. In *Environmental Standards in the European Union in an interdisciplinary Framework*, M. Faure, J. Vervaele, A. Weale, eds. Antwerpen: Maklu, 1994; 197-216.

OECD. *Managing the environment: the role of economic instruments*. Paris: OECD.

Pigou AC. *The economics of welfare*. London: Macmillan, 1920.

Spulber DF. Effluent regulation and long-run optimality. Journal of Environmental Economics and Management 1985; 12: 103-16.

Spulber DF. *Regulation and Markets*, Cambridge: The MIT Press, 1989.

Yohe GW. Comment on polluters' profits and political response. American Economic Review 1976; 66: 981-2.

3 A CARBON/ENERGY TAX FOR SUSTAINABLE DEVELOPMENT

Nadine Gouzée and Stéphane Willems[1]

Abstract

In May 1992, just before the Rio Conference, the European Commission[2] adopted a proposal for a taxation on carbon dioxide emissions and energy. This co-ordinated tax, the first ever proposed to a group of nations, was not subsequently adopted by the Member States of the European Union. Since then, many observers have held that such a project could be considered, like similar initiatives in other countries, to be 'clinically dead'. This paper summarises the reasons why the famous sentence of Mark Twain could be quoted about such a proposal: 'the rumours of my death have been greatly exaggerated'.

The introductory section defines carbon/energy taxation and suggests an analytical framework that is consistent with the concept of sustainable development. The second section describes the evolution of the discussions in international forums from 1992 to 1997. It is argued that the issue of carbon/energy taxation has been continually addressed in the post-Rio process, whether at the level of the European Union, the United Nations Commission on Sustainable Development or the United Nations Framework Convention on Climate Change. The third section of this paper presents the main arguments that have been raised against carbon/energy taxation. It intends to demonstrate that, overall, obstacles can be overcome and that such a project remains critical to the

---

[1] The authors are both members of the Federal Planning Bureau in Belgium. Nadine Gouzée has been the Representative of the Belgian Government to the Commission on Sustainable Development (CSD) and to the United Nations General Assembly Special Session (UNGASS). Stéphane Willems has represented Belgium in the Ad Hoc Group Berlin Mandate (AGBM). The views of the authors are not necessarily those of the Federal Planning Bureau.

[2] Its official name, Commission of the European Communities (CEC), is used in the references.

implementation of the Rio agreements. In the fourth section the paper investigates what can be expected in the future from an institutional perspective.

3.1 Introduction

Since the beginning of the nineties, several countries or groups of countries in the industrialised world have attempted to introduce new taxation systems on energy products. Some of these new taxes were adopted: carbon taxes in Scandinavian countries, carbon/energy and energy taxes in the Netherlands, the energy charge in Belgium. Some did not survive the political process: the energy (BTU) tax proposal in the USA, the 'greenhouse levy' in Australia, the carbon charge in New Zealand and, last but not least, the carbon/energy tax proposal adopted by the European Commission in May 1992, just before the United Nations Conference on Environment and Development, held in Rio de Janeiro in June 1992 (CEC 1992a)[3].

3.1.1 A definition of carbon/energy taxation

These initiatives present different features, but share one common characteristic: the tax is levied on the quantity of the energy produced or consumed, according to an explicit (or implicit) physical measure, which is either the energy content or the carbon content – or both – of the energy products.

In this sense, they differ from traditional forms of energy taxation: VAT regimes, which *inter alia* cover energy products, are levied as a proportion of their (added) market value, whereas existing excise systems usually cover mineral oils only. In fact, these fiscal initiatives have all aimed to create a single, coherent tax regime for a large set of energy products. Whereas carbon taxes cover all fossil fuels (oil, coal, natural gas) according to their carbon content, energy taxes usually cover all forms of energy, – except (new) renewables – according to their energy value. Carbon/energy taxes (as proposed by the European Commission) are a combination of both carbon and energy taxes.

3.1.2 The analytical framework

This article takes the view that these new tax regimes, whatever their differences, can be analysed within the same analytical framework, hence the use of the

[3] For a general description of these tax initiatives, see Baron (1996: 23-34)

generic term 'carbon/energy taxation'[4]. As this paper will make clear, these initiatives can be analysed as different attempts to live up to the challenge of sustainable development. The analytical framework should therefore encompass all dimensions – economic, social, environmental and institutional – of sustainable development. In this sense, it is broader than the theoretical framework of welfare economics, which tends to focus only on carbon taxes as a means to reduce carbon dioxide emissions.

Two other dimensions of sustainable development also need to be highlighted. First, carbon/energy taxation has mainly been proposed to realise very long term objectives. In this context, the slow progress made in adopting proposals for carbon/energy taxation should not itself be an argument against this fiscal instrument. Progress towards sustainable development relies more on the slow evolution of ideas and institutions than on spectacular political decisions. In a sense, the mere fact that there is an ongoing debate on carbon/energy taxation can persuade policy makers, citizens, and businesses to internalise in their decisions the challenges of sustainable development. Hence the importance given in this paper to the debate itself in relevant international forums rather than to the actual decisions taken so far.

Second, instruments like carbon/energy taxation will in the end make sense only if they are applied – and co-ordinated – throughout the world. This highlights the truly global nature of sustainable development. However, it is not reasonable to expect that all countries will implement such an instrument at the same time. It is therefore critical to ensure leadership from industrialised countries, or, at least, a particular group of industrialised countries, like the European Union. In this context, the European Commission proposals have an exemplary value. They represent the first attempt to propose a co-ordinated tax framework to a group of nations. They also represent the political will to assume leadership on this issue, which in turn has fuelled the international debate in other forums. Quite remarkably, in May 1997, five years after Rio, the European Commission presented a revised proposal, in the form of a restructuring of taxation of energy products (CEC 1997). Hence the focus of this paper on these European proposals.

---

[4] The term 'carbon/energy taxation' refers here to all forms of carbon and/or energy taxes. This term is also used in Baron (1996).

3.2 The carbon/energy tax debate in international institutions

Since 1992 the carbon/energy tax has remained firmly on the international political agenda. This is quite remarkable, since it has been the target of heavy criticism from a variety of actors: industry organisations, politicians and government officials, oil producing developing countries, even environmental NGOs. It is particularly true of the debate within the European institutions. The Commission, at the request of Member States, has made successive proposals for a new tax framework on energy products. This has fuelled other debates at the United Nations, which have in turn provided a basis for action within the European Union.

A. The debate in the European Union

3.2.1 The original proposal and its revised form

The debate on the carbon/energy tax began in the European Union after the decision of the Joint Energy/Environment Council in October 1990 to stabilise carbon dioxide emissions in 2000 at the 1990 level and to pursue efforts to improve energy efficiency. In response, the European Commission took on the task of preparing a strategy to limit carbon dioxide emissions and to improve energy efficiency (CEC 1991). After a lively debate within the Community institutions, the Commission adopted in May 1992 a package of measures to be proposed to the Member States. This package of measures included a proposal for a directive on carbon/energy taxation (CEC 1992a). It was clearly intended to show European leadership in the run-up to the Rio Conference.

This tax was regarded as a new, environmental tax, even though its administrative framework was very similar to the existing Community framework for excise duties on mineral oils. It defined minimum tax rates for all non-renewable energy products (including large-scale hydropower). They were set in proportion to the energy content (50%) and the carbon content (50%) of the energy products and would increase from the equivalent of $3/barrel in 1993 to $10/barrel in 2000. Various provisions were proposed to avoid negative economic effects of the tax: conditional reductions or exemptions for energy intensive firms; tax rebates for firms investing in energy saving or carbon dioxide emissions reduction; the so-called 'conditionality clause', which linked the adoption of this proposal to the adoption of similar instruments by the other main industrialised countries; and, finally, 'fiscal neutrality', which meant that the overall fiscal burden should not increase in the Member States.

After several attempts to adopt the proposal in 1993 and 1994, the Essen European Council in 1994 took note of the intention of the Commission to amend its proposal. In 1995, a new proposal was adopted by the Commission: it would provide a common administrative framework, but would no longer set minimum tax rates on energy products (at least in the short term). Member States would be free to decide what would be the proper level of tax rates until the year 2000, when they would need to converge to the level of 10$/barrel. Provisions similar to the ones included in the original proposal were proposed to avoid negative economic consequences. Only the 'conditionality clause' was dropped. Even though it was much more flexible than the original proposal, this revised proposal was no more successful in gaining the unanimous consent of Member States.

3.2.2 The new proposal of March 1997

While many observers considered such a project to be 'clinically dead', a new debate took place within the Community institutions: according to the agreed timetable, the Commission had to propose an update of the directive setting minimum rates on mineral oils. It was a clear opportunity to merge two issues that had been kept separate artificially: the internal market on the one hand, and environment/energy policy on the other. For reasons linked to the internal market, it makes sense to reform the existing excise duty system to include other energy products. Indeed, since most energy products compete with each other, it is logical to set minimum rates, not just for mineral oils, but for natural gas, coal and electricity as well. Such a broadening of the excise duty system makes it possible to pursue different objectives in the field of environment/energy policy.

After a long debate within the Commission, a new proposal was adopted in March 1997. While it looks rather different from the original carbon/energy tax proposal, it resembles it very much in substance. It sets minimum rates for all non-renewable energy products (including large-scale hydropower). New minimum rates would be set for 1998, and would increase in 2000. For 2002, higher rates are proposed as objective rates, but would need to be confirmed (or adapted) in a new Commission proposal to be adopted by Council by January 1st 2001 at the latest. Rates are set separately for each product, but the aim is to have them converge to similar levels (in terms of energy content) at the end of the period considered, i.e. 2002. In fact, objective rates defined for 2002 amount to an (energy) tax of about $5/barrel, which precisely match the energy part of the original proposal, at least for heating fuels[5].

---

[5] For motor fuels, tax rates have traditionally been set at much higher levels.

It is unclear at this stage whether this new proposal will fare better than the two original ones, but it is remarkable that the debate is still going on at European level. The fact that 1997 is the year of the two most important United Nations summits after Rio in the field of sustainable development should in this regard not be considered a pure coincidence.

B. The debate at the United Nations

Although the United Nations agenda has clearly influenced the debate in the European Community, carbon/energy taxation has rarely been an issue for formal discussion in the UN negotiations. This is not surprising, since the worldwide implementation of such an instrument would require a degree of co-ordination between nations that cannot be envisaged at present. Moreover, it is no mystery that some countries, in particular the OPEP countries, lobby very hard to avoid a substantive discussion of such instruments in these international forums. Still, the notions of 'environmentally sound pricing' and 'economic instruments' have been discussed repeatedly, ever since the Rio Conference. This reflects the importance given to the integration of environmental concerns into the process of socio-economic development, which is at the core of sustainable development.

3.2.3 Agenda 21 and the Commission on Sustainable Development

Agenda 21, the action programme for worldwide sustainable development, is the main document that was adopted at the Rio Conference, along with the Climate Change and Biodiversity Conventions. It sets a global framework for action in the environmental, economic, social and institutional domains. Chapter IV of *Agenda 21*, entitled 'Changing consumption patterns', deals with one of the most important 'economic' challenges of sustainable development, that is, 'developing national policies and strategies to encourage changes in unsustainable consumption patterns'. The activities programmed to achieve these changes are, inter alia, 'encouraging greater efficiency in the use of energy and resources' and 'moving towards environmentally sound pricing'. In particular, *Agenda 21* recalls that 'without the stimulus of prices and market signals, that make clear to producers and consumers the environmental cost of the production of energy, materials and natural resources and the generation of wastes, significant changes in the consumption and production patterns seems unlikely to occur in the near future (United Nations 1992a):

The critical question is, however, whether Governments will implement these general recommendations. After Rio, the implementation of *Agenda 21* has not been entirely left to the willingness of governments. A United Nations Commission of Sustainable Development (CSD) was created to review the progress achieved in its implementation. Although environmentally sound pricing is but one of the many topics being discussed, the CSD has taken the time during its sessions to review how various countries or groups of countries have managed to incorporate this notion in their policies as well as to make recommendations in this field.

In 1996, the work of the CSD's Ad Hoc Working Group on Finance and Production and Consumption Patterns was focused on mechanisms and policies such as tax reform and pricing. It was stressed, *inter alia*, that economic instruments can both generate revenue for financing sustainable development and send signals to the market to help change unsustainable consumption and production patterns. The conclusions adopted at the fourth session of the CSD in 1996 recognised the potential of adopting such an approach[6].

In June 1997, the United Nations General Assembly devoted a Special Session (UNGASS), also known as Rio+5, to take stock of the follow-up to the Rio Conference. Its conclusions regarding economic instruments take into account the review undertaken by the CSD: 'in a number of countries, efforts have been made in support of domestic resource mobilisation, including the increasing use of economic instruments to promote sustainable development'. Moreover, the General Assembly recommends in its conclusions that 'governments should consider shifting the burden of taxation on to unsustainable patterns of production and consumption; it is of vital importance to achieve such an internalisation of environmental costs'. (United Nations 1997a)

Five years after Rio, recommendations concerning economic instruments have certainly become more specific. Still, the debate would gain by being focused on more concrete measures. UNGASS has already adopted a more specific conclusion regarding economic instruments in the field of energy. It has recognised the need to internalise progressively environmental externalities in order to achieve a more sustainable use of energy, taking into account the specific economic, social and environmental circumstances of all countries[7]. More important still, UNGASS has decided that the 9th session of the CSD (in 2001) will be devoted to energy issues. This is an important achievement,

[6] Finance is the second important 'economic' challenge of sustainable development. It is treated in Chapter XXXIII of *Agenda 21*, entitled 'Financial Resources and Mechanisms'. In this chapter, developed countries reaffirm, inter alia, their commitment to reach as soon as possible the United Nations target of 0.7% of Gross National Product annually for Official Development Assistance, but they are still far from this target.

[7] It has, however, not accepted the adoption of a proposal from the European Union to recommend the taxation of aircraft fuels, which requires a worlwide implementation.

because it will ensure an overall discussion of energy issues at a global level, and will therefore be a useful complement to the Climate Change negotiations, which only deal with one particular, even though critical, environmental consequence of the use of fossil energy. The issue of carbon/energy taxation will most probably be on the agenda of that session.

3.2.4 The debate in the Climate negotiations

In the nineties, climate change has become a main issue for sustainable development, since it represents a global threat to future generations. In June 1992, at the Rio Conference, more than 150 nations signed the United Nations Framework Convention on Climate Change. As its title says, it is a framework convention: only general commitments are set for all Parties. Basically, they must 'take action' to mitigate climate change and report on these actions. Developed country Parties must take on more specific commitments: they must adopt policies and take measures that 'demonstrate that they are taking the lead in modifying longer-term trends in anthropogenic emissions' (United Nations 1992b: 8). A 'soft' target is set for these countries: the aim is to return greenhouse gas emissions in 2000 to their 1990 levels, but no specific policies or measures are mentioned. Parties have a maximum freedom as regards the policy instruments they wish to implement and no co-ordination of policies and measures is required[8].

However, the importance of the Climate Change Convention has to do with the overall process that it has initiated. First, it has paved the way to a first wave of national programmes to limit greenhouse gas emissions, responsible for climate change. These programmes included many countries' successful or failed attempts to adopt carbon/energy taxes. True, in most developed countries, these national programmes have not proved sufficient to ensure that emissions would return to their 1990 levels in 2000. However, they have provided countries with a lot of experience in the field of Climate Change policy.

Second, it has provided the framework for a continual review of commitments, which will become more specific at each stage. The 'Berlin Mandate' process, initiated in April 1995, is the first of these review processes. It should, *inter alia*, define new commitments for developed countries: targets should be set to ensure significant reductions of greenhouse gas after 2000 and would this time be legally binding; specific policies and measures should also be elaborated.

---

[8] A general reference to the need for coordination is mentioned in art. 4.2.(e)(i): 'Each of the Parties shall coordinate as appropriate with other such Parties, relevant economic and administrative instruments developed to achieve the objective of the Convention' (United Nations 1992b: 9).

In the course of the negotiations, the European Union has tabled a three-tier structure of common, co-ordinated and national measures. Common measures would be measures adopted by all developed country Parties, co-ordinated measures would be measures that should be given high priority by Parties and for which an early co-ordination process should be set up. National measures would be those most appropriate to national circumstances. As a common measure, the EU has proposed applying 'economic instruments to ensure that market prices give appropriate signals to consumers and businesses to limit and reduce emissions of greenhouse gases' (United Nations 1997b: 43). As a co-ordinated measure, it has also proposed a framework for an 'environmental carbon and/or energy tax' that would need to be developed over time (United Nations 1997b: 49).

These proposals are clearly not intended to oblige all developed countries to adopt a carbon/energy tax after Kyoto. This would be an impossible goal at this stage. Rather, they intend to set a binding commitment to use economic instruments as well as to provide an international framework to co-ordinate their implementation at national level[9]. A 'co-ordination process' would be set up within the institutions of the Convention to facilitate such co-ordination. These initiatives need to be understood in a longer term perspective, where it is hoped that policies and measures for sustainable development will be co-ordinated at the global level.

The Berlin Mandate process is now drawing to a close: a Protocol – or other legal instrument – to the Convention should be signed by the Parties in December 1997 in Kyoto. The outcome of these negotiations is still unknown. As they proceed it has become clear, however, that quantified objectives would be at the heart of the negotiations, but it is not yet clear whether they will effectively ensure that significant emission reductions are achieved by developed countries. As regards policies and measures, it is possible that the Kyoto agreement will not even include common or co-ordinated measures, due to the opposition of the United States. However, if legally binding targets for significant reductions of greenhouse gases are eventually adopted, this should provide a powerful incentive to implement carbon/energy taxes in developed countries. Even if the need for worldwide co-ordination is left unanswered, this agreement should trigger new debates on carbon/energy taxation at national or regional level.

---

[9] Developed countries (i.e. 'Annex-1 Countries' in the Convention jargon) have already initiated a process to assess possible candidates for common and co-ordinated policies and measures. This analytical exercise is undertaken by the 'Annex-1 Expert Group on the United Nations Framework Convention on Climate Change', under the secretariat of the OECD/IEA.

3.3 The case in favour of carbon/energy taxation

The international political agenda may well provide many opportunities to discuss carbon/energy taxation in the near future, but it does not mean that there will be any progress on this issue. It is also necessary to show that the obstacles to the adoption of such an instrument can be overcome and that such a project is a necessary step towards sustainable development.

A. A carbon/energy tax for sustainable development

Sustainable development policies, it is commonly argued, need to avoid conflicts between the economic, social, environmental and institutional dimensions of development. A key characteristic of carbon/energy taxation, but also a controversial one, is that it seems compatible with a variety of economic, social and environmental objectives. Another characteristic is that it may also have some advantages for institutional (or administrative) reasons. Even if one particular issue dominates the political debate on carbon/energy taxation at any one time, it is therefore important not to overlook all the other dimensions of sustainable development.

3.3.1 The carbon/energy tax as a price signal

Concern for the global environment dominated the debate on economic instruments, at least in Europe, at the time of the Rio Conference. As a result, carbon/energy taxation has been part of strategies to limit carbon dioxide emissions, the latter being considered to be the main cause of climate change. However, a sustainable management of energy resources is a broader concept. It needs to take into consideration not only global warming, but also other environmental damages, as well as the depletion of non-renewable energy sources. Carbon/energy taxation provides an economic incentive to increase both energy efficiency and the penetration of renewables into the market. These are key features of a sustainable energy path that minimises environmental damage, including climate change, and at the same time slows down the depletion of non-renewable energy resources.

3.3.2 The carbon/energy tax as a source of revenues

Carbon/energy taxation not only provides a price signal to energy markets. It can also generate (quite substantial) fiscal revenues. It can therefore become part of a fiscal policy in pursuit of various objectives. The main purpose of such a policy is to provide governments with a new, alternative form of taxation to finance government spending. The revenues can be recycled in the economy in many ways, depending on the objective that is pursued.

In Europe the carbon/energy tax has mainly been presented as a way to finance a reduction in labour taxes and charges, in order to reduce unemployment. In the US it has been seen as a way to finance investment tax credits or a reduction in corporate taxes in order to boost investments, or, as in the BTU tax debate, to finance a reduction of the public deficit to reduce interest rates. However, there can be many other objectives compatible with sustainable development, like, for instance, financing a reduction of VAT rates on basic necessities for social reasons, financing public spending for energy efficiency programs, or, last but not least, financing development aid, as prescribed by *Agenda 21*.

This 'tax recycling' argument is quite independent of the overall level of taxation. This taxation can be compensated by a reduction of other forms of taxation, in which case there is no increase in the overall fiscal burden[10]. It can also be used to increase the overall level of taxation, if that is considered desirable, or, on the contrary, it can be adopted even though the overall level of taxation is reduced. What is relevant is the fact that it is considered less distortional than – and therefore can replace to a certain extent – traditional forms of taxation, like taxes on income and profits, social security charges and VAT.

3.3.3 The 'multiple dividend' controversy

The multiple dividends that carbon/energy taxation can potentially achieve have given it an exemplary value in the debate on sustainable development. However, it is also what has made it highly suspicious to many economists and policy makers. Economists usually consider that one particular instrument should serve one specific policy objective. They have developed a highly specialised knowledge of each policy field, and, for each one, have selected what they consider the best set of instruments, that is, the set that maximises economic and social welfare. In this context, it is highly improbable that carbon/energy taxation would be selected for a variety of policy objectives. In each case there should be an instrument better suited to the specificity of each policy field. In other words, this 'multiple dividend hypothesis' cannot hold; it is a political, not an economic argument. Policy makers have also been suspicious, however. They are also used

[10] The rather unfortunate term of 'fiscal neutrality' was coined by the European Commission in its original carbon/energy tax proposal. The taxation is certainly not neutral in an economic sense, even though it is compatible with stability of the fiscal burden on the economy.

to working in different spheres, in which a particular set of policy instruments is selected for each policy objective.

That view overlooks the advantage of 'thinking globally', that is, taking an integrated view of development. Obviously, carbon/energy taxation cannot be considered as the best instrument in various policy areas. However, it can be argued that, in many cases, the best instrument in each of these policy fields will either have negative impacts in other policy areas, or be too difficult to implement for administrative or institutional reasons. It is only by highlighting the complexity of the world of 'second bests' that the multiple dividends of instruments like carbon/energy can be illustrated.

However, carbon/energy tax is by no means a miracle solution to the variety of existing economic, social and environmental problems. Each policy field has its own complexity. There is also a virtue in 'acting locally', that is to say, being aware of the details of each specific issue. It is only under specific implementation conditions that carbon/energy taxation will prove beneficial. It needs to be integrated in climate change, energy, environmental, and fiscal policies that include many other instruments. It needs to be designed carefully so as to avoid conflicts of objectives. In fact, it can even be said that there is no such thing as a single instrument called 'carbon/energy tax'. Its existence rests rather on a large set of political decisions that need to be taken at different levels.

The rest of this section presents the different debates around carbon/energy taxation. Since most of these debates, at least in Europe, have taken place in relation to climate change policy, this issue is taken as a starting point, but the overall context of sustainable development is also taken into account.

B. The energy-environment debate

The least controversial aspect of the carbon/energy tax debate has been that it is one of the most effective instruments to limit carbon dioxide emissions. Most observers acknowledge the importance of a price signal that provides a permanent incentive to limit these emissions. However, it is not yet accepted that carbon/energy taxation is truly necessary to achieve significant emission reductions[11]. It is indeed difficult to 'prove' that there is no real alternative to carbon/energy taxation, because there is no empirical analysis that can easily

[11] This section will not discuss in detail the other challenges of a sustainable energy management, that is, the other environmental externalities, as well as the depletion of natural resources. These are very complex issues that cannot be dealt with extensively in this paper. However, it can be argued that carbon/energy taxation would also be needed to tackle these issues.

compare the effectiveness of different instruments to combat climate change. However, arguments in favour of carbon/energy taxation can be presented in a qualitative way.

3.3.4 Economic v 'soft' instruments

Even though economic instruments are considered superior in theory to voluntary, regulatory or other measures, it is also recognised that there is a large potential to reduce emissions cost-effectively at existing prices, i.e. without altering the existing level of prices through economic instruments. As the much quoted statement of the Second Assessment Report of the International Panel on Climate Change (IPCC) says,

> 'Numerous studies have indicated that 10-30% energy efficiency gains above present levels are feasible at negative to zero costs in (the industrial, transportation and commercial/residential building) sectors in many parts of the world through technical conservation measures and improved management practices over the next two or three decades' (IPCC 1995:12).

As a result, it is often argued that these measures should be implemented first, that is, before considering economic instruments. Even though such an economic potential is certainly 'out there', these studies often overlook the administrative costs of government programmes designed to exploit this potential. Even though the measures or practices referred to in the IPCC statement are in principle cost-effective to the energy users, they do not have access to the proper information, nor do they consider energy use as a real priority, particularly in times of low energy prices. Governments need to provide information and technical support to them. They also need to convince them, with various incentives, that such measures and practices should be given priority in their decisions. These programmes can become quite expensive, because they need to 'reach out' to millions of energy users, at least if the ambition of the programmes is to go beyond simple 'pilot projects'. And there is no guarantee of effectiveness, as the experience of countries like the Netherlands, or even the United States, has shown[12].

---

[12] The analysis refers mostly to voluntary programmes, i.e. various programmes that aim to provide targeted information and assistance to energy consumers or to enter into voluntary agreements with industry, because these are the most common form of instruments in developed countries (Willems 1997). Regulatory measures can be much more effective, but their use in climate change policy is limited to highly standardised products, like household appliances, and, even then, administrative costs and institutional obstacles should not be overlooked.

This does not mean that these programmes are useless, but they would be much more effective if coupled with economic instruments, like carbon/energy taxation. These instruments are, indeed, complementary. To take only one example, a carbon/energy tax alone may not be sufficient to convince a consumer to buy the most efficient refrigerator, because he does not have the appropriate technical information. Alone, a labelling scheme indicating the most efficient refrigerator may not be sufficient either, because the consumer does not pay enough attention to his energy bill. Together, a carbon/energy tax and a labelling scheme may at last convince the consumer to change his behaviour.

However, the argument of 'soft measures first' has been used quite often in the carbon/energy tax debate within the European Union. According to some countries, only 'soft' measures are needed to stabilise carbon dioxide emissions in the Community in 2000 relative to 1990. In fact, the European Union itself may manage to stabilise its emissions, but this has to do more with the restructuring of industry in (Eastern) Germany and the shift towards natural gas in Britain rather than to the effectiveness of 'soft' measures. Once the most obvious 'low hanging fruits' have been collected, countries are faced with the compelling reality of rising carbon dioxide emissions. It is indeed the case that most developed countries, except Germany and Britain, will not manage to return their greenhouse gas emissions in 2000 to their 1990 levels.

This issue will soon be raised again, when developed countries will need to take measures to reach the legally-binding targets that should be adopted in Kyoto for the years after 2000. The appropriate mix of policy instruments will obviously depend on the level of greenhouse gas reductions that will be agreed to. With legally binding targets, however, countries will no longer be able to ignore commitments that they have taken up. This may provide the ground for implementing economic instruments in all developed countries.

In the long term, economic instruments will in any case be needed, because existing technologies are not sufficient to reach significant levels of emission reductions. Only economic instruments can provide businesses with the necessary incentives to develop new technologies that are commercially viable. This is perhaps one of the most compelling arguments in favour of the international co-ordination of economic instruments. The implementation of similar economic instruments in all countries will provide the same economic incentive to businesses around the world. This will create a global market for environmentally friendly emerging technologies and will accelerate the process towards sustainable development.

3.3.5 Taxation v emissions trading[13]

In economic theory, emission charges or taxes and emissions trading both possess the intrinsic qualities of economic instruments. The choice between them depends on an evaluation of their overall performance. Once countries have adopted legally-binding targets, emissions trading may actually present a clear advantage. A fiscal instrument cannot ensure that a target is reached because this depends on the price elasticity of demand for energy products, which is not known with certainty. By contrast, an emissions trading system does not by definition lead to over- or undershooting, if all carbon dioxide emitters are allocated emission limits whose sum corresponds to the level of the target. If the system is well designed and if there is appropriate control, a country will reach its agreed target with certainty.

However, emissions trading is best suited to large companies. Faced with an emissions limit, they can easily monitor their emissions and choose what are the most appropriate options to reduce them: invest in technologies that reduce emissions, reorient their activities, or buy emission units on the market. For small emitters like households or small firms, which represent a majority of emissions in a typical country, emissions trading would be very difficult to implement. The administrative costs of allocating and monitoring emissions limits would be too high. Emissions limits could be allocated to intermediaries: energy producers or distributors. However, it would give them an extra market power vis-à-vis energy consumers, since they would be able to raise the price of carbon or energy to 'convince' consumers to reduce their emissions. This may be contradictory to social justice.

The only alternative would therefore be to only give to governments the role of directly influencing directly energy or carbon prices, through carbon/energy taxation. This instrument cannot ensure that an agreed target will be met, but governments are left with the possibility of adjusting tax rates progressively, according to emissions forecasts.

In the end, emissions trading and carbon/energy taxation may also be complementary instruments. Whereas emissions limits could be allocated to large firms, a carbon/energy tax could be levied on small consumers. The debate on the appropriate mix of trading and taxation will certainly start within developed countries when the implementation of the Kyoto agreements becomes part of the political agenda. This will also provide some grounds for the co-ordination of economic instruments, since the implementation of these

[13] An emissions trading system can be designed in many different ways, but these different options will not be investigated here. For a general discussion of international greenhouse gas emission trading, see Mullins, Baron (1996).

instruments throughout the industrialised world will affect the competitiveness of countries. Both instruments can indeed be implemented in such a way as to treat industry favourably or unfavourably, thereby causing possible distorsions of competition.

3.3.6 Energy v carbon taxes

From a strict climate change point of view, carbon taxation is the most efficient instrument, since its tax base is directly related to the emission of carbon dioxide. In order to achieve the maximum effect on carbon dioxide emissions, all existing taxes on energy products should be restructured, so as to set each tax rate according to the carbon content of the fuel[14]. However, in reality, this may conflict with other policy goals that may also need to be taken into consideration.

Prices should reflect the true costs of energy, including costs not accounted for in the production process. Besides climate change, there are, however, other externalities of energy production, transmission and use that may not be accounted for in energy prices, in terms of health and environmental damage or effects on the quality of life. For reasons of economic efficiency, all such external costs should be internalised, possibly through the fiscal system, and not only those of carbon dioxide. This may result in quite different relative prices of energy products than those that would result from merely applying carbon taxation.

The problem with the internalisation of external costs is that there are many uncertainties in their evaluation. In this context, different attitudes may be considered. Research could provide estimates of external costs on which to base taxation policy. Alternatively, rather than waiting for a consensus on the estimation of such costs, in particular those related to nuclear power, the option can be taken of not favouring any source of energy in the tax structure, except those sources for which there is a relative consensus, like renewables. This is one of the reasons why some countries support an energy tax, rather than a carbon

---

[14] In terms of carbon dioxide emission reductions, the difference between carbon and energy taxation may not necessarily be very large, however. The effect of an energy tax on carbon dioxide emissions will depend on its effect on energy efficiency, as well as on the penetration of renewables. In practice, an energy tax may also lead to some fuel substitution, since coal prices will increase the most in percentage terms, as it is the cheapest fuel on a heat-equivalent basis. As for nuclear energy, the choice of increasing or reducing its share in energy supply will in any case be political.

tax[15]. In a sense, they take into account the inability of real world institutions to determine what is the optimal tax structure.

Another reason why an energy tax has been favoured is the competitiveness issue. The countries that are in favour of carbon taxation are those countries, like France, with a low carbon intensity (because of the availability of nuclear or hydropower). If a carbon tax at European level would be adopted, this would give them a large competitive advantage, whereas an energy tax would not change the competitive position of the different countries. That could also lead to some form of energy taxation, if countries would accept the principle that one country should not attempt to gain from levying the tax, at the expense of its neighbours.

This issue of carbon taxation versus energy taxation has been an important obstacle to the adoption of a co-ordinated carbon/energy tax in the European Union. It has, in particular, isolated France in the debate, even though that country would be ready to consider favourably the use of fiscal instruments. Once it is recognised that some form of co-ordinated carbon/energy taxation is truly necessary to achieving significant emission reductions in all countries, the appropriate mix of carbon/energy taxation should be the result of a sound political compromise.

C. The socio-economic debate

Even if a carbon/energy tax is necessary to meet the challenge of achieving significant emission reductions, or, more broadly, of a sustainable energy management, the evaluation of the economic and social cost of that challenge has proved very controversial. Not surprisingly, the notion of 'double' or 'multiple dividend' is at the centre of that controversy.

3.3.7 The macro-economic debate

The question of the macro-economic costs of climate change mitigation is perhaps the most visible part of the scientific debate[16]. Many model results show

---

[15] This argument can be extended to all types of market failure arising from an unsustainable energy management: the externalities of related economic activities, like transport; the monopoly power of energy producers; the 'myopia' of the market as regards the depletion of natural resources in the longer term. If not all market failures are taken into consideration, it is difficult to assess what is the optimal tax structure. It may therefore be better not to change the relative price of (conventional) energy sources.

[16] See Chapters 8 and 9 of the IPCC Report on the Economic and Social Dimensions of Climate Change (Bruce et al. 1996)

that the reduction of carbon dioxide emissions has positive costs (relative to the reference scenario). These costs can be relatively low, when moderate reductions of carbon dioxide emissions are considered. This is usually the case for short and medium term analysis. For instance, most of the global models (general equilibrium models) estimate costs at around 0.5% of GDP for a stabilisation of carbon dioxide emissions in 2010 at their 1990 levels in OECD countries. For larger reductions, costs tend to increase exponentially. This is usually the result shown by long term models that simulate, for instance, the economic impacts of a stabilisation of world carbon dioxide emissions in 2100 compared to 1900. Such a stabilisation in 2100 would be clearly insufficient in light of the long term climate change objectives. In fact, reductions of more than 50% would be needed to stabilise carbon dioxide concentrations, which is the ultimate goal of the Climate Change Convention. Yet, even a stabilisation would mean a strong limitation of emissions compared to the reference scenario and costs would be substantial: around 5 to 10% of world GDP in 2100, according to these studies (Bruce et al. 1996: 336).

According to these studies, costs arise because technological options that bring carbon dioxide emissions reductions are necessarily more costly. If there were no constraint on carbon dioxide emissions, these options (e.g. renewable energy sources) would not appear in the 'optimal' capital stock. This can be challenged in three different ways.

a) Tax recycling issues

These studies usually ignore the importance of tax recycling. However, a decision on tax recycling always follows the decision to levy the tax. No decision is in fact a decision (to use the tax revenue to reduce the budget deficit). Model simulations should therefore always attempt to 'optimise' tax recycling, so as to reduce macroeconomic costs or reveal macroeconomic (or social) benefits. The theoretical issue of tax recycling is quite complex. There will be no attempt to analyse this in detail. However, there are many empirical studies that show positive results of certain recycling options on GDP and employment (Baron 1996; Majocchi 1996; Bossier et al. 1996). For European countries, the effects on GDP are usually not very significant, but the positive effects on employment can be quite large. This is certainly not meant to 'solve' the unemployment problem, but it can contribute to a sound employment policy that combines economic with social objectives. Other tax recycling options also show positive results, like the financing of energy efficiency programmes (Bossier et al. 1996).

b) Synergy between fiscal and non-fiscal measures

The latter option raises an interesting point: the complementarity between fiscal and non-fiscal measures. As already mentioned, non-fiscal measures are best used when they aim to exploit what is called the existing economic, or no-regret, potential[17]. The existence of a no-regret potential at existing prices is usually not assumed in macro-economic models. If a carbon/energy tax can increase the effectiveness of non-fiscal measures, like energy efficiency programmes, in exploiting that potential, their combination may bring benefits that do not usually appear in these models, unless that possibility is explicitly considered in an integrated micro/macro (or top-down/bottom-up) analysis.

c) Technological progress and alternative development paths

In the climate change debate, the two types of benefits mentioned above are 'short term' benefits. In the long term, it is the technological (and human) progress that is key to the issue of costs and benefits of climate change policy. If there is no technical progress, the very significant reductions of carbon dioxide emissions that are needed in the longer term will inevitably cause substantial macro-economic costs. However, in most long term models, technical progress is exogenous. So-called 'backstop' technologies are introduced in the model at a certain date. These technologies, that would emit no or little carbon dioxide, would provide an infinite source of energy, but at a higher cost than conventional technologies. However, technological progress can also be an endogenous process, even though it is very difficult to model. In particular, price incentives, like carbon/energy taxation, can be considered to be key factors in encouraging technical progress in sustainable energy technologies.

More generally, long term models present the impacts of a carbon/energy tax as a deviation from a reference scenario. These models may have a certain degree of flexibility in their response to an exogenous shock, but this shock inevitably represents a welfare loss. However, the long term effects of carbon/energy taxation may also be analysed differently. Combined with other policies that promote sustainable development, carbon/energy taxation may bring society onto a completely different development path with different consumption and production patterns. It becomes very difficult to compare the 'reference' path and the alternative path, as the latter is not merely a deviation from the reference. This issue of different 'sustainable development paths' is a new area of research (Bruce et al. 1996: 281), but it suggests a different perspective on the long term effects of carbon/energy taxation.

3.3.8 The micro-economic debate

There may be a large scientific controversy over the macro-economic effects of carbon/energy taxation, but it is admitted that, even if these costs were positive,

[17] Also called the zero or negative cost potential.

they could be limited by a well-designed tax shift or tax reform, at least for moderate levels of taxation. In particular, prices and wages can be kept under control if the right tax recycling options are chosen and if the reform is accepted socially (CEC 1992b). It is indeed important to 'sell' properly any carbon/energy tax to the public as well as to social partners. If that is done, there should be no fear of a macro-economic shock, like the oil shocks of the seventies.

What matters most to the body politic, however, are the sectoral and distributive effects of a carbon/energy tax. Many groups may well be winners: eco-industry, light industry, services, the average citizen. However, two groups risk being losers: the energy intensive industry and the low income households[18]. Solutions can be found to accommodate these two groups, but it is critical that such solutions be properly designed and implemented. This is perhaps the most important condition for a successful outcome to the carbon/energy tax debate.

a) The trade and competitiveness issue

The energy intensive industry has lobbied very hard at European level to avoid the adoption of the carbon/energy tax proposed by the European Commission. It presented this issue as a trade and competitiveness issue. Competitiveness, however, is an ambiguous notion (Baron and ECON 1997). It is difficult to prove that a whole country will lose its competitiveness and will suffer trade losses from the adoption of a carbon/energy tax. In a typical European country, energy intensive industry (excluding power generation) only accounts for about 20% of carbon dioxide emissions. It accounts for much less in percentage of GDP (about 4 to 5%). In most macro-economic studies, trade effects are indeed not very pronounced.

Of course, this industry may fear that its own competitiveness is at stake, if its competitors do not face similar tax increases. But, even for energy intensive industry, the effect of a carbon/energy taxation on production costs may not be as high as it seems. According to Baron and ECON (1997), the average increase in production costs for energy intensive industry following the introduction of $100 carbon tax is typically about 2% in industrialised countries. These production cost increases may be far less important than, for instance, an increase in the exchange rate. Moreover, competitiveness is determined by many other factors than production costs alone. However, it may still be that one particular firm would face significant cost increases and/or would already be in a difficult position compared to its competitors. It may therefore be forced to cease its operations. Politically, this risk may well be unacceptable.

[18] A third group will always lose out: the energy producers. Energy producers are unevenly distributed across the world. They are not considered in this analysis, although their opposition to significant emission reductions will always be an obstacle to the worlwide adoption of fiscal instruments.

This is one of the reasons why a worldwide co-ordination of fiscal instruments would be needed. However, in the absence of such a co-ordination, carbon/energy taxes, such as the ones already adopted in small European countries or proposed by the European Commission, contain many special provisions for the energy intensive industry: tax rebates for large energy consumers and for firms investing in energy efficiency, exemptions for specific sectors, like the steel, non-ferrous and chemical industry (as in the last Commission proposal). Border tax adjustments have also been proposed in the past to compensate the loss of competitiveness vis-à-vis countries not implementing a similar tax. Even though some small European countries have managed to implement a specific tax system for the energy intensive sector[19], these provisions may well prove to be an obstacle to the adoption of a carbon/energy tax on a larger scale.

There are many reasons for this: they may cause distortions of competition within industry itself; they may also create distortions of competition between countries, if countries implement a different system. The more complex the system is, the more it risks creating new distortions and being too difficult to administer. Last but not least, they usually don't even satisfy industry itself.

There are two solutions to this problem. One possibility is that finance ministries invest in a careful study of the industrial sector and devote appropriate resources to manage a well-designed system of tax rebates and, possibly, exemptions[20]. The other possibility is that the entire energy-intensive industry is exempted, but subject to a system of emissions trading. However, in this case, a new administrative system should also be put in place. In both cases, there should be a clear understanding of the specificity of each industrial sector as well as a well-managed administrative system.

b) The distributional issue

If it is to promote sustainable development, a carbon/energy tax should not increase poverty. That may well be the case, however, if there are no compensatory measures for low income households. Interestingly, the impact of carbon/energy taxation seems to differ across countries. The carbon/energy tax is clearly regressive in Britain, Ireland and the United States, whereas in continental Europe the tax paid would be more proportional to income (Baron 1996: 52). This issue was less prominent in the debate at the European level, even though it may have hardened the position of Britain in the negotiations. It is nevertheless a subject of concern in most countries.

---

[19] See Baron (1996) and Baron and ECON (1997)

[20] Border tax adjustments seem at this stage too difficult to implement and may be contrary to WTO rules.

Fortunately, only a small percentage of households would really suffer from paying the tax. Compensatory measures can be targeted to this income group. As is the case with energy intensive industries, what seems to matter most is the appropriate management of these compensatory measures. Whether in the form of compensation in the income tax system or through insulation programmes for the poor, it needs to be well targeted and well managed. In particular, it may be difficult in highly decentralised countries to divert revenues from the tax (levied by the central government) to pay for social programmes (run by local or regional agencies). A proper co-ordination between different administrations and agencies would in any case be necessary.

3.4 An institutional perspective

There may be many opportunities to discuss issues related to carbon/energy taxation in international forums; this tax may well be necessary for the implementation of Rio's agreements as well as desirable for economic, social, environmental and institutional reasons. Yet, this instrument also has its own set of institutional obstacles. These obstacles may be the most difficult to overcome.

A. The institutional challenges

In many ways, carbon/energy taxation is a new 'institutional territory', because it illustrates all the new challenges of sustainable development: the need to take into account future generations; the need for international co-ordination; the need to combine a global view with a concern for practical implementation issues; the need to integrate economic, social and environmental dimensions of sustainable development. Many conditions are therefore required in order to take the appropriate decisions in the relevant institutions.

First, there should be a sufficient pool of transdisciplinary research to help decision making. As this paper has already shown, many simultaneous decisions need to be taken, if a carbon/energy tax is to be implemented. They all require a different kind of expertise: a fiscal expertise to integrate the carbon/energy tax into the existing excise duty system, an energy/environment expertise to set tax rates at the appropriate level, a macro-economic expertise to design tax recycling options, an industrial expertise to work out the various provisions for energy-intensive sectors and, lastly, a social expertise on the energy needs of low income households, to cite but a few obvious fields of research.

Second, administrations need to broaden their approach to preparing and implementing decisions. In particular, finance ministries need to take on a new, regulatory role. With carbon/energy taxation, they not only collect revenues, which is their traditional role, but also have take into account the economic, social and environmental impacts of taxation. In particular, they need to design appropriate compensatory measures for energy intensive industry as well as for low income households. This is a formidable obstacle, but it is also a chance to improve the co-operation between different administrations and the efficiency of public management in general. These administrative obstacles may well be the reasons why carbon/energy taxation has only been implemented so far by relatively small European countries. In small countries, like Sweden, Denmark, Norway or the Netherlands, that have a tradition of good public management, the administrative complexity of designing an efficient carbon/energy tax system can be kept under control.

Third, at the decision making level, institutions need to show both a sense of leadership and a willingness to co-operate. This is true within a particular country. This is also true at international level, as the carbon/energy tax debate has shown since the Rio Conference, where both issues of leadership and co-ordination are critical to the success of a fiscal strategy for sustainable development.

B. A sense of leadership

At national level, carbon/energy taxation can only have a chance of success if some important political figure or party decides to push it through. At world level, a measure like a carbon/energy tax will never be implemented before having been 'tested' by a (large) group of countries. And it is in the industrialised countries that a step-by-step introduction of a carbon/energy tax is most necessary, because they are by far the largest per capita consumers of energy and resources. Leadership on this issue, however, can only be found so far in the European Union. It cannot be expected from the United States 'whose cultural and economic make-up is linked with unlimited mobility and whose economic philosophy only grudgingly allows for the use of fiscal influences on energy prices' (Von Weizsacker and Jesinghaus 1992).

Only small European countries have so far shown leadership on this issue, together with the European Commission itself. In order to gain momentum, it is necessary that large European countries also show strong leadership at international level. This may happen, if both Germany and Britain remain serious in their willingness to take the lead in climate change policy.

C. A willingness to co-operate

Carbon/energy taxation is perhaps the only policy instrument for which there is both a strong need for co-ordination and a strong resistance against it. At national level, finance ministers have always been very keen on keeping control over matters of taxation. However, carbon/energy taxation makes only sense if it is integrated in a set of policies to promote sustainable development or, more specifically, in a climate change action plan, that also includes voluntary and regulatory measures. At international level, taxation has traditionally been the symbol of the sovereignty of nations. It is therefore a formidable challenge to the notion of international co-ordination that is at the core of sustainable development. The rest of this section investigates what can be expected from the European Union and the United Nations.

3.4.1 The debate within the European Union

It is not surprising that at European level taxation is one of the only policy areas where unanimous consent is required, since it such a symbol of national sovereignty. Whereas most domains of environmental policy require only a qualified majority, environmental taxation needs unanimity. This quite dramatically slows down the transition towards sustainable development.

However, the present state of the Treaty does not mean the end of the debate. History is balancing between two stools. Because they have experimented it for more than two centuries, the US know that the history of Unions goes through phases of expanding federal authority and contracting state authority, as well as phases of contrary winds. The debate about the interpretation of the principle of subsidiarity is only beginning in the EU, whereas the US, after a phase of regulation by the federal authorities, has recently undertaken a general devolution of governmental responsibilities back to the States.

In the short term, a strong pressure towards qualified majority is unlikely. It is most probable that the issues around the Economic and Monetary Union will continue to dominate the political debate in the Union. That may well be an opportunity for the adoption of the new tax proposal of the European Commission, which is precisely meant to provide Member States with a stability of their fiscal revenues in an economic union. It is, however, difficult to say whether unanimity can be found on what remains a controversial issue among Member States.

3.4.2 The debate at the United Nations

In many ways, the decision making process at the United Nations is more open and dynamic than in the European Union.

First, the fact that all countries are represented may be a formidable obstacle to taking decisions, but can also trigger a willingness to co-operate, because countries may also feel that there is less risk of 'free riding'.

Second, the European Union is first and foremost an economic union. National economic interests are traditionally of a higher priority than other objectives pursued by the Union. This is not the case for the United Nations, where the economic objectives are usually 'externalised' in other forums. Of course, one should not be naive, but the terms of the debate are clearly different: in the UN post-Rio process, environment and sustainable development are a priority and the reasons why nations have gathered in the first place.

Third, in the UN negotiation process, the consensus rule must in general be applied, but it is not as rigid as the unanimity rule. In the end, one country cannot simply block the whole negotiating process for fear of driving itself out of the game. A balance is always struck between the widest possible participation and a highest possible level of commitment. A single country cannot just keep the same position forever because it feels protected by the unanimity rule and knows it will remain in the Union whatever happens.

Last, each country must also give an account of its positions and strategies in front of the rest of the World. NGOs are accepted in the negotiating forum, which is not the case in the Council debates. In the open arena of the UN negotiation process, the EU and its Member States delegates may also be encouraged, and prove, to be more dynamic as private persons than behind the closed doors of the Council of Ministers.

The UN negotiating process is, for all these reasons, an interesting one. It allows a stimulating exchange of ideas and can be used to influence decision making in different parts of the world. However, the dynamic nature of this process should not be overestimated. At the level of the UN, there is simply no experience of co-ordination on specific policies and measures, beyond the exchange of information. Countries have very different approaches to policy making, which is less the case within the European Union. Many of them lobby very hard to avoid a discussion on fiscal instruments, like the United States or the OPEP countries.

The main objective in these negotiations would therefore be to create a 'co-ordination process', as has been proposed by the European Union in the climate change negotiations. It would be a learning process where countries could co-ordinate their initiatives to promote sustainable development, create a common understanding and progressively adopt similar policies and measures, like carbon/energy taxation.

3.5 Conclusion

There is every reason to believe that carbon/energy taxation will remain firmly on the political agenda. Indeed, this instrument is critical to the transition towards sustainable development. There are also many reasons to believe that quick decisions should not be expected on this issue. In particular, institutional obstacles remain very strong, whether at national, European or international levels.

However, it can be hoped that the slow progress of ideas in the various institutions concerned with sustainable development will in the end provide an opportunity for a worldwide implementation of carbon/energy taxation. The United Nations provide open forums where such an issue can be discussed at international level, but a strong leadership from the European Union will be needed if this project is to have any chance of success.

References

Baron R. Economic and fiscal instruments: taxation (i.e. carbon/energy). 'Policies and measures for common action'. Working Paper 4. Paris: Annex-1 Expert Group on the UN FCCC (OECD/IEA). 1996.

Baron R. ECON-Energy. Competitiveness Issues Related to Carbon/Energy Taxation. 'Policies and Measures for Common Action'. Working Paper 10. Paris-Oslo: Annex 1 Expert Group on the UN FCCC (OECD/IEA). 1997 (forthcoming).

Bossier F, Bréchet T, Gouzée N, Mertens S, Van den Steen P, Willems S. Politiques et mesures destinées à modifier les tendances des émissions anthropiques de gaz à effet de serre en Belgique. Planning Paper 76. Brussels: Federal Planning Bureau, 1996.

Bruce JP, Lee H, Haites E Eds. Climate Change 1995, Economic and Social dimensions of Climate Change. Cambridge: IPCC, Cambridge University Press. 1996.

Commission of the European Communities. A Community Strategy to limit carbon dioxide emissions and to improve energy efficiency. Communication to the Council (SEC(91)1744 final). Brussels: CEC. 1991.

Commission of the European Communities. Proposal for a Council Directive Introducing a Tax on Carbon Dioxide Emissions and Energy (COM(92)226 final). Brussels: CEC. 1992a.

Commission of the European Communities. The Climate Challenge. European Economy No 51. Brussels: CEC. 1992b.

Commission of the European Communities. Proposal for a Directive Restructuring the Community Framework for the Taxation of Energy Products (COM(97)30 final). Brussels: CEC. 1997.

Gouzée N, Willems S. 'Les accords de Rio: quelle conception de la croissance'. Avec quelle croissance devons-nous apprendre à vivre ? 12ème Congrès des Economistes belges de langue française, Rapport préparatoire. Charleroi: CIFOP, 1996.

Gouzée N, Willems S. Reports from CSD and AGBM negotiations. Internal documents. Brussels: Federal Planning Bureau, 1995-1997.

Gouzée N. Indicators of Sustainable Development: an Institutional Approach. LIst International Conference of Applied Econometrics Association: Econometrics of Environment and Transdisciplinarity, 1996 April 12, Lisbon.

IPCC, IPCC Second Assessment Report, Climate Change 1995, Geneva: PNUE/OMM, 1995.

Majocchi A. Green Fiscal Reform and Employment: A Survey. Environmental and Resource Economics 1996; 8: 375-397.

United Nations. Agenda 21, the United Nations Programme of Action from Rio. New York: United Nations publications, 1992a.

United Nations. Reports by the Chairmen of the Informal Consultations Conducted at the Seventh Session of the Ad Hoc Group on the Berlin Mandate. Note by the Secretariat. FCCC/AGBM/1997/INF.1. Bonn: Secretariat of the UNFCCC. 1997b.

United Nations. United Nations Framework Convention on Climate Change. Geneva: UNEP/WMO. 1992b.

United Nations. United Nations General Assembly Special Session. Conclusions. New York: United Nations. 1997a.

Von Weizsacker E, Jesinghaus, J. Ecological Tax Reform, A policy proposal for sustainable development. London: Zed Books Ltd. 1992.

Willems S. National Programmes for Sustainable Development: a Comparative Study of the US Climate Change Action Plan. Working Paper. Brussels: Federal Planning Bureau. 1997 (forthcoming).

4 THE USE OF NATIONAL SAVINGS IN SUSTAINABILITY ANALYSIS

Anders Christian Hansen
Department of Social Science, Roskilde University, Denmark.

Abstract

One result of economic research on sustainable development is that it is possible to sustain a certain level of per capita consumption indefinitely, even if it is based originally on an exhaustible resource. The condition for this is that the resource rent from the exhaustible resource is invested in reproducible capital. Subsequent research has dealt with properties of depletion rates, shadow prices, and discounting. However, the theoretical results concern a simplified model of a stationary economy and rest on strong assumptions. Using these results defining success criteria, indicators, and targets for actual sustainable development strategies thus requires some modifications of the abstract results. In this chapter, first, the implicit entitlements assigned to each generation are questioned. It is argued that using a savings measure as a success indicator of a strategy for sustainability requires that implicit ethical premises be revealed. In particular it is argued that the savings requirement depends strongly on the growth targets, the demographic changes and the technical progress. Second, the approach is based on the assumption that natural resources are substitutable with reproducible capital, but exactly which resources have this quality has not been subject to much scrutiny. It is argued that only resources that comply with three conditions of substitutability are substitutable, and it is concluded that environmental resources, except for non-unique natural amenities, do not generally meet these conditions. However, if macroeconomic costs of reaching environmental targets can be perceived as liabilities, in the same way as natural resources are perceived as assets, inter-temporal transfer of reproducible capital can easily balance inter-temporal transfer of liabilities.

1. Sustainability and Macroeconomic Analysis

1.1 Reproducible Capital for Natural Resources

A Macro Economic Criterion of Sustainability. In the ten years since the WCED report *Our Common Future* (WCED, 1987) suggested that the goal of 'sustainable development' replaces the goal of 'economic growth' as the superior goal for society, a variety of sustainability indicators and criteria have been suggested and discussed. They describe different aspects of sustainable development, ranging from biophysical over macroeconomic aspects to aspects related to socio-cultural patterns of behaviour. One of the most influential macroeconomic approaches is to define development as sustainable when the total national wealth (including the value of natural resources) does not decline over time. In recent years, a number of empirical studies of changes in national wealth seen from this angle have appeared, notably The World Bank (1995) which builds upon the contributions of Hamilton (1994) and Pearce and Atkinson (1993, 1995).

The empirical studies are based on a theoretical literature of macroeconomic sustainability analysis, which primarily evolved within the framework of the Solow-Hartwick tradition. The core of this literature is based on the line of thinking represented by the Hartwick[1] policy rule of *investing the resource rent* (Hartwick, 1977). Solow (1992) has called it 'an almost practical step toward sustainability' and it links, in an intuitive and theoretically stringent way, economic development to the natural resource basis of the economy.

Governments all over the world have (through The Rio Declaration on Environment and Development and Agenda 21) committed themselves to the following principle: 'To achieve sustainable development and a higher quality of life for all people, states should reduce and eliminate unsustainable patterns of production and consumption and promote appropriate demographic policies'. (Principle 8). All over the world, programs of sustainable development have been adopted on the national as well as on the sector level and in international agreements. Still, it is not clear whether any of the states have been successful in

---

[1] Pearce and Atkinson (1995) noted that J.R. Hicks, J. Page, and R. Solow have contributed as much to the rule as Hartwick has. Therefore they prefer to call it *the constant capital rule*.

their endeavours to live up to their Rio commitments. Do they actually approach a path of sustainable development or are they drifting away from it?

Most of the literature on non-declining wealth has focused on questions concerning how natural resources should be valued in this context and how the maximum sustainable level of consumption can be obtained while using natural capital from exhaustible resources. Less interest has been devoted to questions concerning how much wealth the present generation is actually obliged to pass on to subsequent generations and exactly which natural resources can and should be substituted by reproducible capital. In some of the studies, generations are entitled to the maximum sustainable level of wellbeing or consumption by assumption and it is assumed that there is a substitutable resource that is essential to production. At that level, simplifications and abstractions are needed and it is necessary to make 'heroic assumptions' about substitutability and entitlements in order to get anywhere at all. However, when the sustainability considerations concerning national savings are included in government programs along with other societal goals such as reducing the rate of unemployment and the government debt, then these assumptions become of major importance.

Purpose and Outline of the Chapter. The purpose of this article is to discuss the implicit assumptions behind the idea of a non-declining national wealth as a sustainability criterion and to suggest some ways to make the assumptions less 'heroic' and more in accordance with country specific sustainable development goals. First I discuss the ethical premises on which the idea is built. What does a generation actually owe to the next generation or, to put it the other way round, what is it entitled to from its parent generation? The second part deals with the question of substitutability. Exactly which natural resources are substitutable and which are not? In the third part I suggest some extensions of the savings criterion for macroeconomic sustainability. I do not, however, intend to add to the already large body of mathematically heavy literature using optimal control theory to define and derive properties of the model economy. The discussion of assumptions can be done with a very modest use of mathematics. First, however, I shall introduce some experiences from the use of this kind of criteria so far and a general accounting framework for the discussion.

1.2 The Accounting Framework

Green National Accounting. The non-declining national wealth approach is related to the *green GDP* approach by accounting for at least part of the income from natural resources as *sales of assets*. Revenue from sales of assets is not really income and should rather be treated as *capital consumption* or *depreciation.* In this sense, income earned by the use of natural resources often reduces national wealth.

It may seem inappropriate to use the national accounting framework when analysing an issue such as sustainable development, which is of a global nature. Sustainable development on the global level, however, cannot be separated from sustainable development on the national level and it is the responsibility of the close to 200 national governments of the world to bring their nations on a sustainable development path.

Accounting for natural resource consumption requires a greening of the *standard accounting model.* In the traditional accounting model, natural resources do not play a role, except for the categorisation of industries. The model basically consists of a wellbeing function and a production function. The wellbeing function is supposed to be representative for all the households in the economy and the per capita consumption is the only argument in the function. The output of the aggregate production function is a composite good using capital and labour as inputs. Production and consumption are linked together by a static condition of aggregate supply and demand being in equilibrium and of dynamic identities, which determine the capital stock and the labour force.

This framework is obviously too narrow for analysing sustainability since the natural resource basis of the economy is assumed away. However, only a few simple extensions are needed to make the standard model greener. The greening of the standard macro model has evolved in the theoretical literature by adding natural resources as arguments in the wellbeing function and in the production function and by specifying dynamic identities for each of the natural resource stocks in question.[2] The constraints on natural resource stocks and the way they enter into the functions are central in analysing economies where the natural resource basis rather than accumulated capital and labour has become the limiting factor of production. As will be argued below, it is not enough to include natural resources as inputs in the functions. Physically, they represent a system of throughputs and the sustainable scale of the throughputs can be limited from the emission side as well as from the input side. Daly (1996) refers to this as the scale problem. In this chapter I shall use the term throughput to describe the individually distinct flows of separate qualities and quantities rather than an aggregate flow of matter and energy measured in, e.g., tons and joules.

[2] See, for instance, extensions of the production function in Solow (1974), Dasgupta and Heal (1979), Hartwick (1977), of the wellbeing function in Krautkramer (1985), and of both in Dasgupta (1995) and Hartwick (1990)).

The greening of the standard model requires accountable definitions of natural resources and a standard taxonomy of the natural resource basis can be summarised in the following three categories:

- exhaustible resources (mineral reserves),
- renewable resources (fisheries, forests, arable land), and
- environmental resources (dispersion and deposition services, general and human specific life and health support functions, recreational services, biodiversity, and amenities)[3].

The ozone layer, the atmospheric permeability for heat radiation, clean air, soil and drinking water resources can be classified as life and health support functions. Renewable energy resources like hydro-, wind-, sun- or geothermal power are left out because energy consumption from such sources do not affect subsequent generations negatively.

Natural Capital Consumption. The point of greening the standard model is that to the extent these natural resource categories enter the production and wellbeing functions, they can be expressed in terms of reproducible capital and treated analytically in much the same way.

In such an economy, we can define net national wealth as the total of the stock of fixed and inventory capital, net foreign assets, and natural resources valued as capital. Non-declining national wealth then means that the total of human, man-made, and natural capital remains constant or increases over time. Leaving human capital out of the analysis, the gross national savings must be at least as large as the total consumption of natural and man-made capital to keep national wealth constant.

In national accounting terms, the sustainability criterion for one accounting period can be defined as

$$S_t \geq S_t^* = D_{Kt} + D_{Nt} \qquad (4.1)$$

where S_t is gross savings, S_t^* is the sustainability requirement, D_{Kt} is the consumption of fixed capital and D_{Nt} is the consumption of natural capital. Consumption of fixed capital is straightforwardly obtained from national accounts. Natural capital consumption can be defined as the consumed units of

[3] Compare with, for instance, Hartwick and Olewiler (1986) for an economic approach and Groot (1992) for an environmental science approach.

natural resources multiplied by their respective shadow prices or, more specifically,

$$D_{Nt} = (\mathbf{R}_{t-1} - \mathbf{R}_t)\mathbf{q}_t + \mathbf{Z}_t\mathbf{v}_t, \tag{4.2}$$

where $\mathbf{R}_t$ is a row vector of the various stocks of natural resources, $\mathbf{q}_t$ is a column vector of the unit rents[4] associated with the individual types of natural resources, $\mathbf{v}_t$ is a vector of damage costs per unit of emitted pollutant, and $\mathbf{Z}_t$ is a row vector of emissions.

The World Bank (1995) includes exhaustible and renewable resources (R_t) such as fossil fuels, some non-fossil minerals, and reserves of tropical timber in the accounts. Emissions of carbon dioxide are used as an – admittedly crude – indicator of environmental damage (Z_t). The values attached to exhaustible and renewable resource consumption (q_t) are the unit rents of the resources in question, crudely estimated as 50% of the market price of the extracted or harvested raw material. Carbon dioxide emissions are assigned a value (v_t) of \$20 per metric ton of carbon.

1.3 Preliminary Exercises

Insufficient Savings. Accounting for natural capital consumption makes a difference. One of the first analyses of savings net of natural capital consumption by Pearce and Atkinson (1993, 1995) showed that a range of countries had positive net savings according to traditional national accounting, but turned out to have negative savings when adjusted for resource consumption.

The World Bank (1995) computed the *genuine savings* of the major developing regions of the World. Genuine savings are defined as the national savings in excess of the total consumption of reproducible and natural capital. The diagram below shows for each region, the ratio of genuine savings to GNP. Curves above the zero-line represent national savings that exceed the consumption of reproducible and natural capital. When curves are below the zero-line, the genuine savings rates are negative; that is, gross savings are insufficient to cover the consumption of reproducible and natural capital.

[4] Only a part of the resource rent is capital consumption if it is assumed that relative prices will increase due to increasing scarcity. This is because capital gains in that case will add to the resource as a future asset.

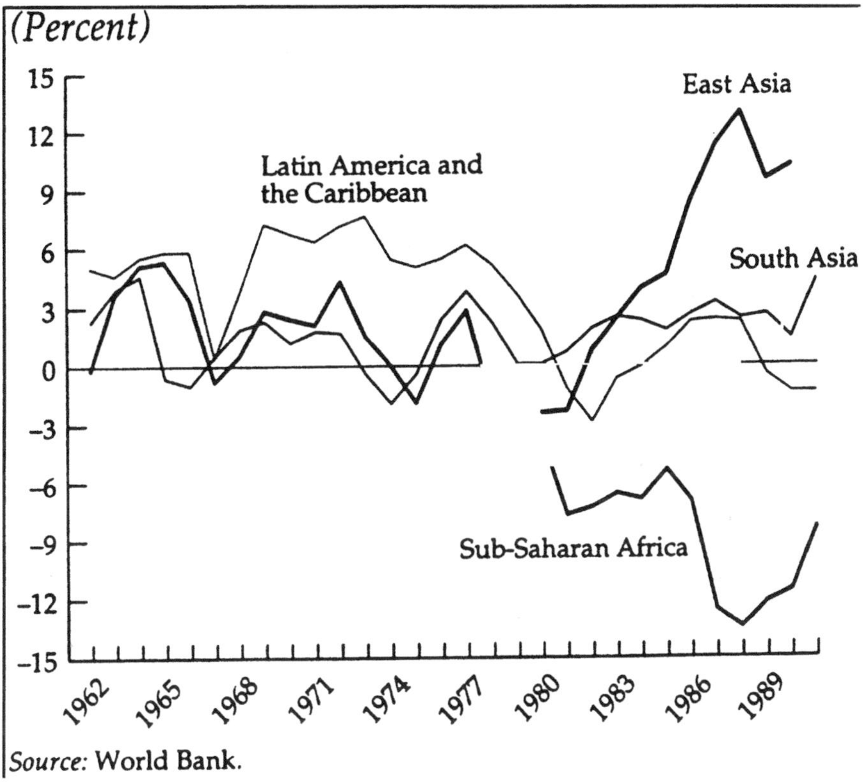

Figure 4.1. Genuine Savings as a Share of GNP, by Region, 1962-91

Source: Reproduced from The World Bank (1995).

The diagram very clearly reveals the potentials of the *genuine savings rate* as an indicator of sustainability. South Asia (in the 80s) and Latin America and the Caribbean (in the 70s) have over the recent decades been able to save a little more of their national income than they consumed. The genuine savings of East Asia and Sub-Saharan Africa in the 80s are mirror images of each other. At the one extreme East Asia shows a tremendous and increasing ability to accumulate more capital than is consumed. At the other extreme, the economies of Sub-Saharan Africa throughout the 80s have steadily consumed more of their stocks than has been added to them. Despite the fact that the average standard of living is already alarmingly low in this region, the stocks by which future generations should produce their standard of living seem to be *systematically undermined* by the present generations.

Policy Recommendations. These exercises demonstrate that accounting for natural resource consumption may provide new valuable information that has previously been overlooked. This information is about a kind over-consumption that is not captured by the standard indicators used in macroeconomic analysis such as the current account deficit, the budget deficit and the rate of inflation, which usually indicate over-consumption. However, since the idea behind the concept is to treat natural resources similarly to manmade capital, it can only indicate that there is a problem, not identify where in the economy or its resource basis it lies.

In fact, the *policy recommendations*, which may be derived from that kind of analysis, do not primarily concern natural resource policies. The problem of insufficient savings frequently or perhaps even usually lies in macroeconomic variables such as gross investments or the propensity to consume (see, e.g., World Bank 1995; p. 55). The standard recommendation in such cases would be to use macroeconomic instruments to achieve macroeconomic goals such as budget balance and current accounts balance. In the case of Sub-Saharan Africa, the negative genuine savings are hardly unaffected by decades of wars, civil wars, natural disasters, lack of diversification of agricultural production, and counterproductive government regulation. These problems cannot be remedied by using macroeconomic instruments nor by natural resource policies but rather through institutional reform and other than economic policies.

The fundamental policy recommendation is that genuine savings at least should be non-negative. What does a *positive* rate of net savings indicate? Pearce and Atkinson (1995) use the criterion "for determining whether a country is on or off a sustainable development path" (p. 168). In other contexts it is referred to as a sustainability test that is weak in the sense that if a development fails the test, it is definitely unsustainable, while a development which is unsustainable according to other criteria might pass this weak test. In other words, as a necessary but not a sufficient condition of sustainability. This also seems to be how The World Bank (1995) perceives it when stating that "where genuine saving is negative, it is a clear indicator of unsustainability" (p. 53). It does not claim that a positive rate of genuine savings indicates sustainability.

Which rate of genuine savings should a government that seriously tries to get the economy on a sustainable path then aim for? How should a savings measure enter into long term strategies for sustainable development? Should the ratio of genuine savings to GNP be zero, a specific number, or just as high as possible? Which is better? And how is this concept of economic sustainability confined to and linked with concepts of ecological sustainability? I will address these questions by discussing the entitlements that implicitly are assigned to generations when a savings rule is used as sustainability criterion and by discussing under which conditions natural resources are substitutable by reproducible capital in the greened accounting model. These questions are intertwined with the questions of defining a better measure of measuring wellbeing than per capita GNP (see, e.g., Nordhaus and Tobin 1972; Daly and Cobb 1989). My intention is, however, to keep clear of that discussion in order to keep the size of the chapter down.

2.The Entitlement Problem

2.1 Ethical Premises

The implicit ethical principle behind *non-declining wealth* is that each generation is *obliged* to pass on at least the same stock of general wealth as it received from its parent generation. If each generation is so obliged, then each generation must also be *entitled* to the same general wealth as its parent generation. The principle of non-declining national wealth is commonly perceived as an interpretation of the often cited Brundtland definition of sustainable development as a "development that meets the needs of the present without compromising the ability of future generations to meet their own needs" (WCED 1987; p. 43). This quotation, however, only includes the part of the Brundtland sustainability definition that concerns intergenerational equity. The second half of the definition is as important to the commission and it concerns primarily the question of equity between rich and poor regions of the world. If the intention is to quantify sustainable development in a way that reflects the international consensus, then this should be kept in mind.

According to welfare theory we can determine an *efficient allocation* of the goods in an economy based on consumer preferences. If we could eliminate all market failures and imperfections we would be able to obtain it by letting consumers express their preferences on competitive markets. A *just distribution*, however, can neither be determined nor obtained in a similar fashion. It is a matter of ethics, politics, and power. But any distribution that the society prefers is, in principle, consistent with an efficient allocation obtained through appropriate use of markets.

The closest we may come to a rational determination of distribution is probably the Rawlsian suggestion of a maximin criterion (Rawls 1972). According to this criterion, the distribution, which is most in favour of the least well off, should be preferred. The difference between this distribution criterion and strict egalitarianism is that inequality is compatible with it to the extent that it improves the prospects of the least well off and the consumption opportunities for the society as whole.

While Rawls himself was reluctant to use this criterion on intergenerational distribution problems, Solow (1974) adopted it as an ethical premise of sustainability. Solow did, however, recognise that any economy that would live up to such a principle in intergenerational distribution would be a stationary economy. Thus, it is of little use to, e.g., developing economies that aim for development from low income to middle income economies in a sustainable way. An alternative approach could be to recognise the actual long term development goals of governments and development organisations, which in some respects reflect the entitlements of the future that are considered just by the present.

2.2 Optimal or Sustainable Savings?

One of the simplest entitlements one can think of is that a generation should at least be entitled to make the most out of its GNP. That is, divide it between consumption and investment (savings, consumption in the future) in such a way that it maximises the wellbeing of the generation over time. In a longer perspective, matters are not that simple, however.

The question of how much a nation should save is not new to economics. Saving is the consumption that is abandoned by the present in order to increase consumption in the future and the economic problem is how to determine the right balance between present and future consumption. Finding the savings rate that maximises the present value of present and future consumption can solve the problem. Present value, however, involves discounting and discounting means that future consumption is assigned less weight than present consumption. The more distant the future and the larger the discount rate, the lesser weight is given to future consumption. Thus, an optimal growth path, which is derived using a constant rate of discount, will under certain conditions eventually to lead zero consumption. Such a growth path might be preferred by the present generation but would certainly not be fair to future generations. Ramsey (1928), who was a pioneer in this field, found that discounting in this context "is ethically indefensible and arises merely from the weakness of the imagination." (p. 261).

The present generations might even prefer an 'overshoot and collapse' path to a path with modest but stable growth if maximisation of present value is the only criterion. An 'overshoot and collapse' path is characterised by booming growth followed by a drastic drop in output, such as several of the scenarios that the Club of Rome warned against in the early 70s. While there is not a simple answer to the question of how much a nation should save, it is obvious that what is optimal for the present generations is not always what would be preferred by the future generations. The focus on *sustainable development* as an alternative to *optimal growth* can be interpreted as a growing recognition of this.

Following an optimal growth path, which tends towards zero consumption in a distant future would, according to Chichilnisky (1996), be the dictatorship of the present generations over the future generations. If, on the other hand, a growth path were chosen which ignored the rate of discount, it would be the dictatorship of the future generations over the present. Then a present generation could not maximise consumption or utility over its lifetime. Thus, Chichilnisky (1996) proposes a principle of *non-dictatorship* as a basis for a macroeconomic sustainability criterion. The *Chichilnisky sustainability criterion* combines discounting of the near future with not discounting of the remote future. Applying this criterion, the socially preferred growth path will be *optimal* in the long run and *overtaking* in the *very* long run. An overtaking growth path is a growth path where output or consumption exceeds all other growth paths from a certain point in time. Some of the obvious difficulties with this criterion are how to distinguish between the long run and the very long run and which weights to assign to the future generations.

Norgaard and Howarth (1991) hold a similar view when analysing the relation between the rate of discount and the rights assigned to each generation in an overlapping generation model. They argue that the dispute in the field of project appraisal of how to choose a rate of discount which at the time could ensure efficiency and sustainability only arise because economists have forgotten how distributional equity and allocational efficiency interrelate. They conclude that sustainability cannot be addressed through efficiency but through assignments of rights to each generation. The discount rate is the rate of return on foregone consumption in order to secure future consumption. And in this sense it is relevant for inter-temporal maximisation of the present generation's wellbeing. Moreover, the discount rate so defined *depends* on the assignment of rights to generations.

The empirical exercises of the World Bank (1995) and Pearce and Atkinson (1995) implicitly assign such entitlements to generations. They do not intend to determine the savings requirement necessary to achieve optimal growth, but only to find a minimum requirement, which at least should be fulfilled. That is, they do not intend to make the criterion serve as a criterion of inter-temporal efficiency too.

The requirement that national wealth should be *non-declining* implies that a generation is entitled to as much wealth as its parent generation and is obliged to pass on the same amount to the next generation. The requirement has the advantage that it excludes the possibility of preferring an overshoot-and-collapse growth path to a modest-but-stable growth path, even where the present value of the former exceeds that of the latter.

2.3 Total or Per Capita Wealth?

In accounting terms, a non-declining wealth criterion implies that the next generations are entitled to the same *total* national wealth as is to the disposal of the present generations. This premise is also worth considering. The needs of future generations naturally depend on how large they are.

If, for instance, we return to the genuine savings rates of African economies, it would indeed be a major progress if they were raised at least to zero. This would make as much wealth available in the future as is available today. But it should be considered that the African population is expected to double from 1995 to 2025 (WRI 1995). Then a constant stock of national wealth would mean that the average African in 2025 would have *only half* the wealth at his disposal as the average African in 1995. This would be highly unsustainable according to almost any conceivable sustainability definition and the sustainability indicator should reflect this. Then, it would be the *per capita* rather than the total wealth that would be sustained by complying with the sustainability criterion. In the event, this is also a standard approach when analysing economic growth.

It would be sensible to take other demographic trends into account too. In many OECD countries, a fall in the share of labour force in population is anticipated. It makes a case for additional savings if a given standard of living has to be maintained with a smaller labour force. On the other hand, the labour force itself is not necessarily the limiting factor when there is high unemployment and low capacity utilisation. Then output can, in principle, be increased without a proportional increase in the capital stock.

2.4 Wealth, GNP, Consumption or Wellbeing?

In a seminal paper, Solow (1974) defined intergenerational equity not in terms of national wealth but in terms of consumption. As long as we assume zero technical progress it does not make any difference, but as soon as we approach real life economies it makes a great difference.

Why should future generations be entitled to a certain stock of national wealth? Wealth is not an end in itself, but a means to produce useful output. Measuring needs in terms of per capita consumption seems more relevant than measuring it terms of wealth. It makes a considerable difference whether we use non-declining wealth or non-declining consumption. If we assume Hicks-neutral technical progress of 1.5% annually, then the generations living about 50 years from now will be able to produce and consume more than twice as much as we do on the same capital stock. It might be good for them, but are they *entitled* to it? The answer will probably depend on where the question is asked. In Vietnam, it would be: yes of course; they have the right to several times as much. In Denmark, however, the answer would possibly be no.

The principle is also controversial the other way round. The standard approach to empirical analysis of technical progress is to calculate total factor productivity (TFP) as the residual economic growth that cannot be explained by growth of inputs of labour and capital. Estimates of TFP growth show that 18 out of 22 African countries experienced *negative* TFP growth rates over the recent three decades (Nehru and Dhareshwar 1994). This result was independent of whether human capital was modelled as an independent factor of production or not. A steady downward trend in productivity is not sustainable in any sense. The 'development' of economies with steadily declining productivity would be unsustainable even if they had positive rates of genuine savings.

The study also showed that human capital, measured as formal education, had a considerable impact on growth rates. The results provide a case for including human capital in national wealth as suggested by Pearce and Atkinson (1993) and Hamilton (1994). Additionally, it would have the effect of equipping the wealth concept with the beauty of completeness since we cannot think of anything on earth which is not human, man-made, or natural. Hamilton (1994) suggests accounting for education expenditures as investments. However, education expenditures are only a small fraction of human capital investments. According to the theory of human capital (e.g., Mincer 1974), most of the investments in human capital are the foregone earnings of students, apprentices, and workers while they improve their skills. Moreover, according to the Nehru and Dhareswhar study, the potential future earnings due to increased enrolment rates are highly uncertain. For the sake of reliability of the savings rule, it might be better to include the average TFP growth over the recent years as it represents *realised* rather than hypothetical enhancement of the productive capacity of the resources. In that case, it must be assumed that the productivity increase is permanent. For economies with negative TFP growth, it would raise the recommended minimum savings rate. For countries with positive TFP growth rates, it would lower it.

If the wellbeing of future generations is our concern, an argument parallel to the TFP growth argument can be made in the wellbeing function. Then a given level of wellbeing can be maintained with less consumption, that is, the part of the consumption expenditures that has been labelled 'defensive expenditure'. They include, e.g., defence expenditure and environmental protection expenditure. If the pressure that makes this kind of spending necessary could be relieved, then they could be reduced without reducing the overall level of wellbeing.

2.5 Growing or Non-declining?

In a simple growth model, a non-declining capital stock means steady state growth. Neo-classical growth theory has introduced the question of which steady state, we should aim at (Phelps 1961). If the growth strategy follows the golden rule (Do unto others as you would have them do unto you), we should aim at the steady state with the highest possible per capita output. This steady state corresponds to the overtaking path above.

In most of the world (the developing countries and countries with economies in transition), this is obviously not the present state. Thus, developing strategies for such countries typically aim for high rates of growth of the capital stock.

Using a sustainability criterion of non-declining capital stock (such as zero genuine savings) would be a strike in the air if sustainable development requires a considerable growth in the capital stock.

As pointed out by Solow (1974), economic theory can analyse the existence and properties of the maximum per capita consumption growth. However, it cannot tell us how us how to get there.

Alternatively, the planned growth rate for the economy could be interpreted as an indication of how fast a given country wants to approach the golden rule level of the capital stock. On this basis, it is possible to estimate the savings required for sustainable development as it is defined for that particular economy.

2.6 Non-Declining Resource Stocks?

The question of what future generations are entitled to in terms of natural resources is at the heart of the sustainable development debate. A frequently used distinction is the distinction between *weak* and *strong* sustainability (see Turner 1993 for a survey).

The weak criteria can further be divided into *weak* (excepting 'critical capital', i.e., non-substitutable resources) and *very weak* (all resources can be substituted) and the strong ones into *strong* (natural resources can be substituted with each other) and *very strong* (no substitution at all) criteria. The distinction reflects differing views on whether future generations are entitled to the natural resources in the physical quantities and qualities that the present generations have found them in or in their capacity of generalised value, of capital.

The rule of non-declining national wealth is definitely a weak criterion. But given some substitutability between natural resources and other goods, there is simply no reason for assigning less than a right to the optimal level of environmental quality to each generation, even if the present and preceding generations have run the environmental qualities down to sub-optimal levels. This perspective is already contained in government environmental programmes all over the world aimed at restoring environmental quality rather than just maintaining the current poor quality. It is a normally acknowledged part of the concept of sustainable development. For instance, The President's Council on Sustainable Development states that sustainable development is also to "preserve and, where possible, to restore the integrity of natural systems" (cited from Daly 1996; p. 13).

Moreover, there can be considerable time lags involved in environmental regulation, notably in greenhouse gas regulation. Providing the optimal level of environmental quality in 2050 is thus a task for the climate policy around 2000. If the future is entitled to a certain level of environmental quality, then there is a limited *budget* of environmental pressure to which the economy has to adapt. This adaptation to ecologically sustainable patterns of consumption and production entails costs that are negative entitlements (obligations) and that must enter the inter-generational capital accounts as liabilities. Accounting for such obligations gives rise to two new questions. First, how should an international obligation, such as the obligation to reduce greenhouse gases, be distributed among countries and, second, what does it cost to keep the budget?

2.7 Entitlements and Sustainability Criteria

When non-declining national wealth is used as a criterion of sustainability, it implicitly is based on the ethical premise that future generations are entitled to at least the same stock of national wealth as we have to our disposal. In country specific analyses of sustainable development it is not necessarily the relevant ethical premise. The discussion above suggests, first, that the needs of the future generations should rather be defined in terms of *per capita consumption and environmental quality* than in terms of national wealth.

Second, the savings requirements needed to keep the national wealth from declining are those needed to replicate the current state of the economy. These requirements are obviously irrelevant to a country, which, e.g., plans to advance from a low-income to a medium-income economy. One alternative is to derive the savings requirements from *actual* goals of sustainable development with respect to per capita consumption and environmental quality. The trade-off between the two is in many instances a problem that must be solved by the preceding generations.

3. The Substitutability Problem

3.1 The Importance of the Problem

The standard objection to a savings rule as a sustainability criterion is: 'How can investments in more highways compensate future generations for our poisoning their water?' This question embraces the problem of substitutability in the production function as well as the problem of substitutability in the wellbeing function. The former reflects the everyday notion that more capital in use *ceteris paribus* means more pollution. The latter problem concerns to what extent future generations will be as happy with more consumption goods as with a clean environment.

Dealing with these issues in the formulation of a sustainability criterion makes the task a lot more complex, but it is important as the resulting savings requirements differ considerably according to the assumptions of substitutability, while academic research can live with (and on) 'heroic assumptions' strategic orientation.

It is also an indispensable criterion for getting meaningful balances out of any accounting exercise that the accounts are complete. This is also true in green national accounting. If the value of resources of unknown magnitude is omitted from the analysis, the figures of required net saving would be of little use. This is even more important when one is accounting for growth or discovery of natural resources as additions to the natural capital since the error from incompleteness in that case can go both ways. Therefore, one must dedine exactly which resources can be substituted and thus included in the accounts.

In the literature on natural capital the question of substitutability is repeatedly emphasised. Pearce and Atkinson (1995) advocate the idea of accounting partially for critical, i.e. non-substitutable, capital, but are reluctant to point out which categories of natural resources should be treated as such. Solow (1992) admits that "the claim that a feature of the environment is irreplaceable, that is, not open to substitution by something equivalent but different, can be contested in any particular case, but no doubt it is sometimes true" (p. 21). However, studies contesting the actual substitutability of the particular cases are more than rare.

It could be argued that these problems cannot be solved within the realm of economics. It is technological, biophysical, and ethical problems that must be solved by the respective disciplines. It is probably true to some extent but still, we can get much closer with economic reasoning.

The question of substitutability can be seen as a question of how the natural resources enter either the production function or the wellbeing function. Solow

(1986) argues for using the substitutability: "The current generation does not especially owe to its successors a share of this or that particular resource. If it owes anything, it owes generalised productive capacity or, even more generally, access to a certain standard of living or level of consumption. Whether productive capacity should be transmitted across generations in the form of mineral deposits or capital equipment or technological knowledge is more a matter of efficiency than of equity. " This clear statement is, however, immediately modified to a stronger one by: "The preservation of natural beauty is a different matter since that is more a question of direct consumption than of instrumental productive capacity" (p. 142).

I shall make use of this distinction between substitutability in the production function and substitutability in the wellbeing function in the section on substitutability below.

The question addressed in each of the following subsections, then, is under what conditions reproducible capital can take the roles of the natural resources in the economy, that is in the production function and the wellbeing function. There can be three types of substitutability conditions:

- *The technological condition,* which concerns substitutability in the production function.
- *The preference ordering condition,* which is about the shape of the wellbeing function.
- *The appropriability condition* concerning the interaction between the functions.

3.2 The Technological Condition

The Roles of Natural Resources in Production. Natural resources that enter the production function do so mainly in two ways. First, they serve as a basis for primary industries and part of the income created in those industries comes from the resources. The resources do not in themselves produce income, but because access to their use is restricted, the producers can earn a *differential resource rent*[5]. Second, they serve as inventories or providers of particular raw materials that are essential to the global economy.

With respect to the role as contributor to GNP, natural resources can easily be substituted by reproducible capital. The net present value of all future receipts

[5] See note 4.

net of extraction costs of optimal extraction is in principle the value of the access to extraction when discounted by the rate of return to invested capital.

It is much more difficult to state the extent to which reproducible capital can replace natural resources as inventories of specific raw materials. This question must be posed on the global level since all raw materials are traded on international markets. The provision of raw materials is of importance to all national economies irrespective of the origin of the raw materials.

Renewables. Raw materials from renewable resources like fish stocks, forests, and arable land enter the production function in their capacity of rent sources and material throughputs. One should distinguish between resources that serve as suppliers to an international market and resources that serve as food providers for the local community. In the former case there is in principle a rent to the economy, but it is often competed away due to over-utilisation. In the latter case, there are obvious limits to technological substitution by capital. In contrast to exhaustible resources, they also enter the wellbeing function in several ways, ranging from the global concern for rain forests and whales to the local concerns for amenities.

Exhaustibles. In the literature on substitutability, the main focus has been on exhaustible resources because of the irreversible character of their consumption. The possible exhaustion of these resources gives rise to concerns about not only marginal substitutability but also whether they are fully replaceable by reproducible capital.

Dasgupta (1993) argues that, for analytical purposes, resource augmenting technical progress can be reduced to substitution of raw materials by reproducible capital. Resource augmenting technical progress includes the discovery of new ways of economic use of the materials, increased productivity or more cost effective methods in the extraction process, increased efficiency in the use of resources, use of lower grades of deposits, and recycling of the materials. All of these types of resource augmentation can be reduced to substitution of vanishing resources by reproducible capital.

Exhaustible resources can be divided into fuel and non-fuel minerals. Fuel minerals include fossil and nuclear fuels and non-fuel minerals include metals and other raw materials. It is a widely held view that only *fuel minerals* are the *ultimate limited resource* since other materials can be substituted or extracted in abundance if sufficient energy inputs are available (e.g., Dasgupta 1993). In the economic literature the substitution of fossil fuels is handled by discussing the properties of a constant elasticity production function with an exhaustible resource.

The Elasticity of Substitution. Solow (1974) provides a framework for the economic analysis of resource substitution in the production function. In terms of a production function with a constant elasticity of substitution, the substitutability can be described by considering three special cases. The elasticity of substitution, e, is the rate of change in factor composition (e.g., reproducible capital / raw material) due to the rate of change of their relative prices. Consider the CES production function (used by, e.g., Dasgupta and Heal 1979) in which output is produced by the vital raw material, z, and reproducible capital, k, where both are normalised by labour inputs, L.

$$f(k, z) = \{\alpha_1 k^{(e-1)/e} + \alpha_2 z^{(e-1)/e} + (1-\alpha_1-\alpha_2)^{e/(e-1)}\} \qquad (4.3)$$

where $\alpha_1, \alpha_2, (1-\alpha_1-\alpha_2) > 0$ and $e > 0$.

The case where the constant elasticity of substitution, e, is *less than one* results in the long run in total exhaustion of the resource and thus an end to production. Consequently, the only sustainable level of consumption of z in this case is zero consumption. In order to postpone the time of human extinction, a strong sustainability criterion would be required, distributing the vital resource over as long a period as possible.

The case where the elasticity is *exactly one* (the Cobb-Douglas form) leads to ever increasing marginal productivity (and price) of the resource that will continue to enter the production function in decreasing amounts for ever. An additional requirement is that the elasticity of reproducible capital should exceed the elasticity of natural capital.

The case where the elasticity is *larger than one* means that there is no major problem in substituting the raw material by other inputs. Production can easily take place without the particular raw material.

Testing Substitutability. If we can reduce the question to a matter of whether the elasticity of substitution is less than, equal to, or greater than one, it is seemingly an empirical question whether exhaustible resources are substitutable by reproducible capital.

Dasgupta and Heal (1979) go thoroughly through the various possibilities and conclude that we might very well live in a world where the elasticity of substitution is at least one. Referring to comparisons of the factor shares in gross income they estimate that the elasticity of exhaustible resources is perhaps four times as large as the elasticity of reproducible capital. Add to this that at least some technological progress must be expected in the future. Dasgupta and Heal conclude that, "even in the absence of any technical progress, exhaustible resources do not pose a fundamental problem..." (p. 205).

It would, however, be a mistake to assume that the technology, and thus the elasticity of substitution, at, say, the end of the 21st century will be identical to the technology in middle of the 20th century. Infinite substitution as in the Cobb-Douglas case or with constant and large resource augmenting progress is a theoretical but not necessarily a practical opportunity. Dasgupta and Heal (1979) note that the elasticities which can be observed on present data might be different from those that would appear by large values of the capital-resource ratio, which cannot be avoided in the very long run. Hartwick (1989) explains more colourfully about infinite substitution, that "the physics of this phenomenon is unclear since teaspoons of say oil will be followed by eyedropperfulls and ultimately fractions of molecules followed by fractions of atoms" (p. 36). In the very long run, an elasticity of substitution of exactly one is thus not necessarily enough to ensure substitutability. Victor (1991) argues additionally that the production and operation of capital itself requires resource inputs that may be less substitutable than in other types of production. Heal and Chichilnisky (1991) find empirical evidence that could support the hypothesis of such a complementarity between capital and energy inputs.

Then we are back to the statement that only resources for which $e > 1$ at the end of the exhaustion period do not pose a fundamental problem. But this is somewhat tautological because it means that only the resources are substitutable for which a substitute technology will exist at the time of exhaustion.

Economic Feasibility. It is pretty much a matter of belief in technology whether this is the case or not for fuel exhaustibles. Aage (1984) argues that economics can hardly give the answer to this question and recommends a precautionary principle. Authors with an optimistic view of technology typically use the argument that human history is one long chain of substitutions of basic raw materials (see, e.g. Dasgupta 1993). Why should this not continue in the future? Opponents as Aage (1984) argue that when there is basic uncertainty[6], rational decisions cannot be made on the assumption that such forecasts become true.

While it is hard to imagine that *energy* could be totally replaced by the other factors of production, it is no news that the fossil *fuels* we rely on today can be replaced. Substitutes already exist in the form of other mineral fuels such as lower grade fossil fuels and nuclear fuels. At another level substitutes for mineral fuels exist as well. They include renewable energy sources such as bio

[6] Or *scientific* uncertainty cf. the subsection on the preference ordering condition below.

fuel, solar, wind, geothermal, and wave energy. Finally, partial substitutes for energy exist in the form of energy saving technologies.

These technologies are called backstop technologies as they allow substitutes to eventually take over the role of exhaustibles. They are already *technically feasible* but at a much higher cost than fossil fuel based energy. If energy prices increase as the existing mineral fuel reserves are depleted, the backstop technologies will eventually become competitive or *economically feasible*. The problem is the price at which this happens. As noted by Dasgupta and Heal (1979), under the assumption of a constant rate of discount the price increases will be insignificant in most of the depletion period. Thus, developers of backstop technologies will be likely to hold back the backstop technology until a price level is reached which can compensate the innovation costs. Then, Aage (1984) continues, the scarcity induced price increase may occur shortly before ultimate exhaustion. To this we can add an increasing concentration on the supply side of the fuel markets. For instance, only eight countries had a ratio of oil reserves to production above 40 in 1995 (BP, 1997). This scenario implies a risk that energy will become prohibitively expensive to the poor regions in the World at some point of time. It could cause a considerable drop in the standard of living for future generations in high-income countries too. If such predictions hold, mineral fuels are only substitutable in practice if the backstop technologies become economically feasible at a pace that matches the exhaustion of the mineral fuels.

This brings economic analysis back into answering the question of substitutability. It is a fundamental economic observation, that production costs decline over time due to *dynamic economies of scale*[7]. It is not the passage of time itself, but the experience that is accumulated through the use of the technologies that cause costs to decline. Then, the backstop technologies can only be available to all future generations at, say, the fossil energy cost at the present if the present generations use them despite their higher costs at the present. And then, substitutability of fuel minerals is a question of whether we use the alternatives on a sufficient scale.

The Throughput Paradigm and Emission Budgets. In the 80s the concerns about energy resources shifted from focusing on energy resources as inputs to energy resources as throughputs of matter and energy. The use of fossil fuel reserves is now considered limited by the capacity of the atmosphere to absorb carbon dioxide rather than by the finite mineral fuel reserves. The absorption capacity of the atmosphere is, however, not only a physical fact but as much a

[7] Also referred to in microeconomic theory as learning economies or the Horndal effect.

political decision of the risks of climate change that we are prepared to impose on the future generations. If, for instance, it is an anthropogene temperature rise of a certain maximum it follows that there is a limited global carbon budget. This budget is likely to be smaller than the carbon stored in the remaining fossil fuel reserves.

If it is assumed that the world does not keep the carbon budget we are back to the original substitutability problem. Then the purpose of analysing sustainability becomes somewhat more dubious since the development is then by definition not sustainable with respect to one of the most important features of sustainable development.

If, however, it is assumed that the world will eventually keep the carbon budget over the century, then there is also a total obligation of investments in emission reductions. This is analytically easier to handle since investments in emission reductions are actual substitution of fossil fuel by reproducible capital in the form of the backstop technologies mentioned above. How these investment obligations should be distributed over time is as much a question of intergenerational distribution as a question of efficiency. However, if investments in emission reductions are postponed, this imposes investment obligations on the future and it will – at least within some limits – leave the future as well off as the present if such postponements are followed by a corresponding increase in savings. In this way financing of the adaptation to the carbon budget is substitutable.

Similar considerations can be made for the budget of nuclear fuel and plants that eventually becomes nuclear waste which now in many European countries are considered in excess of the budget.

Marginal substitution of throughputs without economic disaster is something that happens all the time as relative prices change. This does not prove that any throughput scale can be reduced to its sustainable level without serious economic consequences. The difference between the costs of the first and the last emission reductions of an emission reduction program can be quite large. This is illustrated by several cost analyses of carbon dioxide emission reduction options, revealing low and even negative cost options that obviously should be used first. Also cost estimates from computable general equilibrium models tend to come up with very modest costs in terms of foregone consumption opportunities. Whether the last emission reductions can be undertaken without severe reduction of the standard of living is unknown, however. The theoretical warnings of complementarity, of low elasticity of substitution by much lower throughput intensities, and of myopic markets tell us that it might not be possible. The present generations do, however, affect the cost level of the last emission reductions through the directions that technological development takes today.

Conclusions. To sum up this subsection, natural resources play two roles in the production function of the economy. One as a source of national income (resource rent) and one as a source of throughputs that are essential to the global economy. Reproducible capital can, without major obstacles, replace other sources of national income. The real substitutability problem concerns the role as throughput and not its marginal but rather its full or almost full substitutability.

If future generations are entitled to a certain level of per capita consumption and certain levels of environmental quality at the same time, this requires that a set of not only technically but also economically feasible options are available at that time. This requirement can be met by implementing an inter-temporal throughput budget including development of substitution options. In other words, substitution in the future is not given but decided upon by the present. The crucial condition is that the inter-temporal budget is eventually kept. Combined with the requirement of non-declining per capita consumption it will typically require that the costs of keeping it be distributed more or less evenly over the entire budget period.

3.3 The Preference Ordering Condition

Zero or Optimal Pollution. The question of substitutability of natural resources in the wellbeing function is basic in environmental policy and to the conflict between environmentalists demanding 'no pollution' and economists demanding 'optimal pollution'. As noted above, nobody denies that there are limits to marginalism in the case of deciding on the preferred level of environmental quality. Even sworn neo-classical economists accept that it is necessary to distinguish between environmental qualities that are 'irreplaceable' and those that are not and between 'critical' and substitutable resources.

Environmental Quality Targets. In this section, it is argued that environmental qualities typically belong to both categories. They are irreplaceable at low levels and marginally replaceable at high levels. Some examples of environmental policy might help to illustrate this.

One example is that a concentration level of atmospheric chlorine of 2 ppm is regarded as the critical level above which an ozone hole can occur (Miljø- og Energiministeriet 1995). The wellbeing effects of a hole in the ozone layer are uncertain but the target of regulation is set at the critical level. On the other hand, a more demanding level could have been chosen. Before the emergence of anthropogene atmospheric chlorine the level was as low as 0.6 ppm (Miljø- og Energiministeriet 1995) and this level of environmental quality would probably be preferred if it could be attained at no cost. The choice of target indicates that there is limit beyond which it simply is unacceptable to expose future generations to these risks.

Targets for concentrations of atmospheric carbon dioxide are not set internationally at the time of writing this article, but the Intergovernmental Panel for Climate Change (IPCC) has provided a series of alternatives ranging from 450 ppm to 750 ppm. There is immense scientific uncertainty about how high the concentration can increase before the resulting radiative forcing becomes seriously damaging. Thus, if the world community decides upon a target for carbon dioxide concentrations and thus implicitly a global emission budget, it can only be based on precaution regarding the unknown but potentially catastrophic events that could occur beyond the limit. Again an even lower concentration ratio would probably be preferred, if it could be obtained at no cost.

Similar patterns can be found in European regulation of acidification where *critical loads* play a role of benchmarks for target setting and in numerous other environmental regulation issues. The target is typically set at some minimum quality below which the environmental quality is unacceptable and above which it is better but also more costly. In the following, I will put this into a theoretical frame, which can help defining substitutability and non-substitutability of environmental resources.

Lexicographic Preferences. The examples can be generalised to other natural resources that enter the wellbeing function, that is, the various types of environmental quality. They include life and health supporting functions, amenities ('recreational and aesthetic functions'), and ethical functions ('existence value'). The wellbeing function transforms these attributes of environmental quality in conjunction with private and public consumption goods to health, safety, satisfaction of material needs, and other aspects of wellbeing.

Preferences for environmental functions such as life and health must be downwards bounded before the point at which life ceases or becomes unbearable. In other words, substitution stops at some lower limit where the environmental good becomes *immeasurably* more important than consumption goods. The level is independent of income or the budget restriction. This kind of preference ordering is called *lexicographic* preferences (see, e.g., Malinvaud 1976 for details).

Also the health and life of other species are to some extent subject to lexicographic preferences. On a survey about the value of biodiversity, reported by Spash and Hanley (1995), it could be concluded that a substantial fraction of the persons asked had lexicographic preferences for the existence of various species and ecosystems. In these cases, the environmental good has an *intrinsic or existence value* that can neither be measured nor substituted. Apart from this study, empirical evidence of the nature of environmental preferences is rare. A reason for this could be that the preference ordering is usually an assumption rather than a study object.

The region of lexicographic preferences is usually excluded from textbook expositions of welfare aspects of environmental choices, probably because the relevance of marginal trade-offs is rather limited in these cases. It may, however, be useful to consider the wellbeing function of future generations using the full picture of their preferences. The diagram below shows how the indifference map of the representative individual of future generations might look if the indifference curves are downwards bounded by lexicographic preferences for the environmental good.

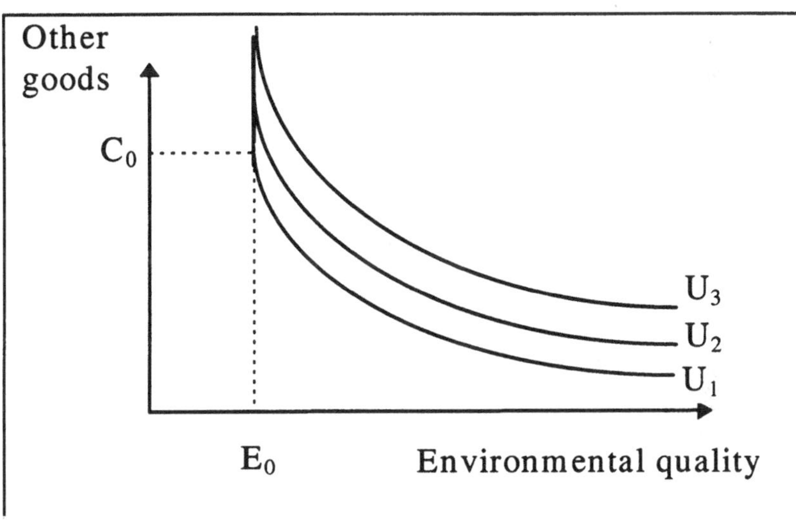

Figure 4.2. Indifference Map Bounded by Lexicographic Preferences

What the figure shows is that below a certain level, E_0, the value of the environmental good becomes infinite. A fully lexicographic indifference curve would be the straight line from (E_0,C_0) and upwards. For environmental quality levels higher than E_0, there is marginal substitutability between environmental quality and other goods. At environmental qualities in this region, it makes perfect sense to estimate the willingness to accept a reduced environmental quality and to compensate for it by reproducible capital. For many environmental problems, the choice is, however, of a different nature, since the prospects are that environmental qualities will be below E_0.

Uncertainty, Irreversibility, and Option Value. The indifference map of the representative individual in Figure 4.2 requires perfect information about the wellbeing effects of the environmental quality. It also requires that substitution along the indifference curves can go both ways. However, the process of consumption of environmental goods often possesses the characteristics of uncertainty and irreversibility.

Arrow and Fisher (1974) found that an *option value* must be taken into account in the presence of uncertainty and irreversibility. The opportunity of undoing the original decision, should it turn out to be disadvantageous, has a value itself, the option value. Irreversibility means that environmental resources, such as pristine natural areas that are sacrificed for economic purposes, such as a dam, cannot be re-established, or only at prohibitive cost. Arrow and Fisher showed that even under the assumption of risk neutrality, the combination of irreversibility and uncertainty makes it rational to preserve more nature than would have been the case if only expected net benefits had been considered.

Another factor behind lexicographic preferences could be what Funtowicz and Ravetz (1991) call the decision stake, which is about the scale of the risk involved. It should make a difference whether the decision stake is the health of a single person or extinction of the all humanity. Particularly the combination of irreversibility and high stakes may be a source of lexicographic preferences.

The type of uncertainty considered by Arrow and Fisher referred to a sequence of decisions of sacrificing environmental resources for economic aims. As the process continues more information about the net benefits is gained, which may lead to estimates of net benefits that differ from the original estimates. According to Randall (1991), the option value can be integrated in the computation of the expected value of a policy. The computation requires a complete listing of all the possible states of the environment, all the probabilities attached to each of them, and the values of each state. This kind of uncertainty can be divided into *supply* and *demand uncertainty*. Supply uncertainty arises from the uncertain effects of the alternative policies and demand uncertainty from the policy independent but uncertain shifts in demand that may arise.

A more difficult type of uncertainty is the *scientific uncertainty* or, in the environmental policy context, our ignorance about ecosystems, how they work and interact, and what they mean to the environment on which we depend. If the possible states of the environment are not known, an expected value cannot be calculated. The concept of *quasi-option value* implies a similar complication since it involves uncertainties of the values of each state due to new discoveries, e.g., of new economic uses for genetic material.

The combination of unknown risks of the possible states and unknown risks of their values is what characterises many environmental problems and it gives rise to

The Precautionary Principle. Considering the presence of irreversibility and uncertainty gives optimal environmental policy a bias towards more environmental preservation. It makes up the prime case for a *precautionary principle* in environmental regulation that has become the guiding principle in a number of international and national environmental policies.

The socially preferable minimum level of environmental quality will typically be somewhere to the right of E if uncertainty and irreversibility are involved and a precautionary principle is applied. If levels below the minimum level are ethically unacceptable, then the target level itself becomes sort of a lexicographic boundary. We define a collective indifference map of the representative individual to reflect not only his preferences as a consumer but also as a citizen. As a consumer, he weighs individual consumption against individual risks. As a citizen, he has ethically determined boundaries for the risks (or rights to environmental quality). There are many examples that we take more precaution at the collective or political level than on the individual. Numerous security regulations concerning our behaviour in the traffic, at work, or even at home are examples of this. Individuals that want to take risks that are unacceptable to society are not always allowed to do.

The figure below shows how a typical collective indifference curve is likely to look.

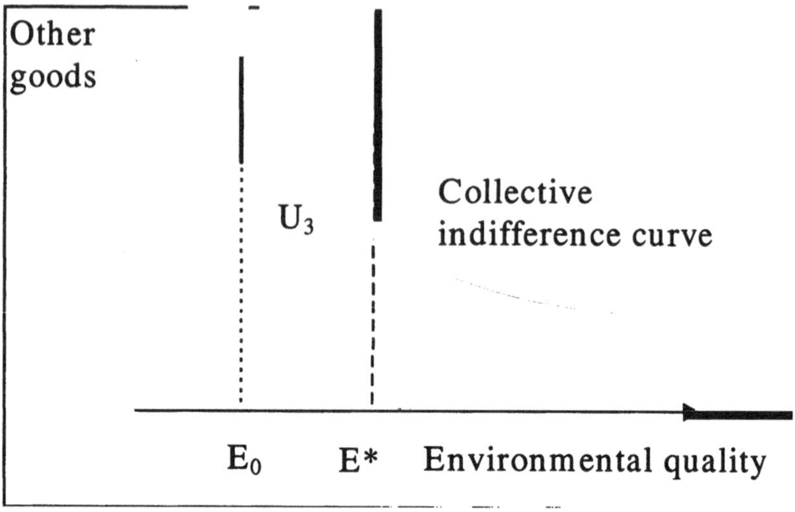

Figure 4.3. The Collective Indifference Curve of the Environment-Consumption Trade-Off

Figure 4.3 shows the minimum acceptable level E* at a precautionary distance from E_0. The collective indifference curve that is relevant for long term environmental decisions is the likely indifference curve for the representative individual bounded by this level. For simplicity of exposition, only one of them is shown in Figure 4.3.

Given collective indifference curves as in Figure 4.3, the optimal level of environmental quality can be at E* or even above if the costs of conversion are sufficiently small. The question of substitutability depends on the point where you are located. If future generations have an environmental quality above E*, it is to a certain extent substitutable by reproducible capital. If it is below or equal to E*, only reproducible capital can be substituted by improved environmental quality, not the opposite way. What implications do the presence of a lexicographic region on the collective indifference map have for the maximisation of the wellbeing of the representative individual in, say, 2050? This question can be answered by identifying the highest possible indifference curve within the production frontier. Obviously, regardless of the shape of the production frontier, as long as it is concave, the maximum wellbeing will be at an environmental quality at or above E*. In some cases, the lags involved in changes of the environment are of such a length that the environmental quality that affects a given generation cannot be changed by this generation. This is to a large extent the case for atmospheric concentration of greenhouse gases. This means that the production frontier is a straight vertical line. However, it can be more to the right or to the left depending on emission regulation between 2000 and 2050. Its vertical extension depends on the size of the national wealth and the technical progress in the period. Therefore, from the point of view of 2000, these possible production frontiers together build a concave production frontier with a trade-off between environmental quality and other goods that makes marginal substitution to the right of E* meaningful.

This can also be seen as a way to minimise the catastrophic events, the probability occurrence of which are insignificantly small but nevertheless believable and the outcome of which is unacceptable. This combination is discussed by Perrings (1991) who finds that it is rational to apply the precautionary principle under these assumptions.[8]

Period and Profile of Emission Reduction. This adds flexibility to the optimal solution of the balance between consumable output and environmental quality in the future. When the target *level* is set, the costs in terms of consumable output in the future depends on, first, when the target level is planned to be reached and, second, the profile of regulation costs over the period. Depending on the choice of emission reduction period and emission reduction profile almost any figure of emission reduction costs could be true, and the real trade of between environmental quality and consumable output is in many cases a question of the target year rather than the target level.

Conclusion. The conclusion is that environmental quality is marginally substitutable by reproducible capital, when it is above a certain politically determined precautionary level. The assumption underlying this conclusion is that the preference map that guides political decisions about the environment is divided into a lexicographic and a substitutable region. When the environmental quality of the present is lower or is projected to decline below this level with current practices then the first relevant economic question is how and how fast to approach this level. The costs of industrial conversion to cleaner technologies and of conversion to cleaner consumption patterns depend on how fast the changes should happen. In this sense not only the target level but also the year by which the target level is fulfilled is a matter of substitution between environmental quality and other goods, even when environmental quality is in the lexicographic region.

---

[8] The assumptions are labeled 'focus loss' and 'reserved rationality'.

3.4 The Appropriability Condition

Property and Use Rights. The appropriability condition is a question of *property or use rights* to the natural resources and the capital by which it is substituted. In this sense it is a matter of institutions. In an abstract and closed economy, the size of the total capital stock available to the economy is of importance to the representative consumer who consumes its product. Similarly, the quantity of the natural resources that are appropriated by the agents and the government in this economy affects the wellbeing of the representative consumer. In such a model, it makes good sense to assume trade-offs between natural resources for reproducible capital between the generations. The same average consumer that would have benefited from the natural resource if it had been preserved appropriates the returns from the accumulated capital.

Time Horizon. However, not all natural resources share the appropriability characteristics that are typical to reproducible capital. Some pollutants evaporate within the period in which they are emitted. This is, for instance, the case for particle, noise, and smell pollution. These types of pollution are not as durable as reproducible capital. In that case it does not make sense to compensate for environmental utility loss by capital. The beneficiaries of the investments are different from the losers of environmental utility. Only if the damage itself is lasting, does it make sense to compensate *future* generations for it. For instance, if a continuing noise pollution causes certain species to revert to other habitats and, thus, irreversibly destroys a unique ecosystem. Another example could be if the continuing environmental pressure caused lasting health effects but these could of course only be indirectly disadvantageous to future generations. The *type* of investment does, however, affect the environmental quality with respect to short-lived pollutants. Investments in cleaner production and transport certainly give future generations environmental quality for reproducible capital, while investments in polluting technology and transport system do not.

Redistributing Institutions. The Hicks-Kaldor criterion is sometimes used as a gedankenexperiment on the international or global level, although it still is subject to debate among economists. The Hicks-Kaldor criterion considers it socially beneficial if the outcome of a change is that the total gains exceed the total losses when the losers are not actually compensated by the winners. However, within nations or at least within welfare states there is a case for using the Hicks-Kaldor criterion: redistributing institutions exist and they will more or less ensure that some of the economic gains eventually will benefit the losers. But comparable institutions do not exist between nations. There is no guarantee that economic gains in one part of the world will be redistributed so that it will benefit losers elsewhere in the world economy. Charity exists indeed as a permanent phenomenon in the international economy but it cannot be compared to the redistributing institutions of a welfare state.

Recognising this, national savings cannot compensate for environmental damage in other countries. If the environmental quality entering the wellbeing function of economy A is degraded due to pollution from economy B, the representative consumer in economy A would not benefit from increased capital accumulation in economy B.

There are plenty of examples of global and cross-border environmental problems: emissions of greenhouse gasses and ozone depleting substances, acidification, eutrophication, and other sea pollution. Biodiversity is also primarily a global concern. Even some local pollution of particles, nutrients, pesticides, heavy metals, and other harmful substances is known to eventually disperse to transnational environments such as the sea.

Sustaining the wellbeing of future generations of a specific nation thus requires that the losses of natural resources are compensated by accumulation of reproducible capital in *that particular nation* and not in the nations where the emissions originated. Actually doing this would, however, place a disproportionate burden on the shoulders of the developing countries (that still pollute less than the industrialised countries) in order to secure the World's future. Thus, it has been argued, an indicator of sustainability ought to respect the 'polluter pays principle'. However, a normative principle cannot without further notice be transformed into a positive measure like that. Government policies ought to follow the 'polluter pays principle' in environmental regulation but it would be purposeless also to do it in national accounting when the polluter in fact only pays himself.

National Interests. There are other appropriability problems associated with resources that play a strategic role in the national economy, such as oil reserves often do.

First, in many small open economies, developing, and transition economies, primary production also serves as an important source of *foreign exchange*. There are not very many industrial branches that can replace oil extraction in this respect. Thus, countries that enjoy comfortable export earnings from natural resource exploitation and want to share this with the future generations should be concerned about the sum of the natural resource capital and their foreign assets. For instance, it could be a sustainability target to maintain this total at the present or another specified level.

Second, in many instances it is considered important to have national *control of the supply* of specific materials, notably fossil fuel reserves representing essential inputs in the production process. To ensure that future generations can have the same level of supply security, alternative domestic energy sources can be built up as, e.g., fossil fuel reserves are exhausted. This is, for instance, the approach of the Danish government which aims for a rate of self sufficiency in primary energy of approximately 40% when the domestic oil and gas reserves are exhausted (Miljø- og Energiministeriet 1996).

Conclusions. To sum up, the appropriability of wealth stocks or of their services matters for whether they are substitutable by each other. Only resource stocks the appropriability of which matches that of the reproducible capital stock at the aggregate level are substitutable by nationally accumulated reproducible capital.

3.5 The Substitutable Resources

Three Substitutability Conditions. Against the background of the above discussion, it can be concluded that a natural resource, to be substitutable with nationally accumulated reproducible capital, must comply with all of the following three substitutability conditions.

A) The *technological* conditions: The key condition is that, for each environmental quality that future generations are entitled to, there is a credible inter-temporal throughput budget reducing the flows of throughputs to appropriate scales. Credibility includes development of feasible options and distribution of the economic burden over the period.

B) The *preference ordering* condition: The expected future level of environmental quality must exceed the lexicographic boundary.

C) The *appropriability* condition: The national appropriability of the services of the natural resource consumed must correspond to the national appropriability of the reproducible capital accumulated.

The inter-temporal throughput budget is equivalent to a programme for emission reduction. If a throughput budget is given, the emission reduction targets can be derived and *vice versa*. Assuming linearity in emission reductions, the emission or throughput budget is

$$B = (Z_0 + Z_T)T/2 \qquad (4.4)$$

where T is the target year, Z_0 is the throughput flow at time 0, and Z_T is the target level.

If the budget, B, over the period, t=0;T, is given rather than the target level, then the annual emission budget in the period will be B/T and the target level

$$Z_T = (B - Z_0)2/T \tag{4.5}$$

Which Natural Resources are Substitutable? Materials, the throughputs of which are limited from the emission side *are* being substituted by reproducible capital if they follow the optimal emission reduction program. They comply with B without further notice. They comply with C in their capacity of production factors with nationally restricted access. In their capacities of providers of domestic primary energy supply and of sources of foreign exchange, they are substitutable by particular forms of reproducible capital: renewable energy production facilities and foreign exchange earning or saving capital.

Arable land, forests, and fish stocks comply with A in their capacity of factors of production, but not necessarily with B in their capacity of parts of or habitat for ecosystems and species.

Many environmental resources, however, do not comply with the other two conditions for substitutability. They typically represent public goods on a global or trans-border level that are not matched by corresponding redistributing institutions. Preferences for environmental resources do frequently have lexicographic boundaries.

The 'net-consumption' of the absorption capacity of ozone layer depleting substances and greenhouse gasses do neither comply with B because of their expected level below the lexicographic boundaries nor with C because of their global nature. Some emissions causing acidification, tropospheric ozone, urban air pollution, eutrophication, toxic substances in the environment, and noise pollution do not comply with C because of their trans-border nature or their short-livedness . In some countries they neither comply with B because the health and life risks they are expected to induce are collectively unacceptable.

One would expect the preferences for environmental amenities (the recreational and aesthetic functions) to be somewhat different since beauty is not a necessity for survival. In most countries they are within the substitutable region of environmental quality. On the real estate market, environmental amenities are traded daily against other qualities of housing location such as proximity of job and service centres. In countries where environmental impact assessment is mandatory in public physical planning it is even elevated to the collective level. However, nature areas of particular beauty or uniqueness are often preserved without notice of the potential economic benefits of alternative uses, simply because they are considered as a part of the national heritage which future generations are entitled to. That is, they are at, or below, the lexicographic boundary. In this sense, the indifference map above could represent environmental amenities, as well. Therefore, aesthetic qualities of nature, such as beautiful landscapes and other amenities comply with all of the conditions unless they are unique to an extent that make them considered as national or even global heritage.

The extinction of some species, ecosystems, and genes in general or at specific locations is level unacceptable on the collective. Consequently, these resources do not comply with B. To the extent that biodiversity is a global concern, it neither complies with C.

Conclusions. Paradoxically, we end up with resources with instrumental productive capacity *and* natural beauty as the substitutable natural resources.

4. Some Extensions of the Savings Rule

Obligations of Emission Reduction Investments. A standard inter-temporal emission budget must specify the profile of the financing or obtaining of the programme. Assuming linearity in emission reductions, the annualised emission reduction target of this conversion process for each year could be

$$AR_t = (Z_0 - Z_T)/T \tag{4.6}$$

where T is the target year and Z_T is the target level.

Clearly, a postponement of the conversion process to subsequent years implies a reduction of future consumption opportunities. To ensure a non-declining standard-of-living, and provided that the target level and target year is not changed, this will have to be compensated by increased savings.

Postponed conversion can be defined as

$$PC_t = AR_t - (Z_{t-1} - Z_t) \tag{4.7}$$

and the cost per reduced emission unit of the entire emission reduction programme can be defined as

$$b_t = CC_t/(Z_t - Z_T) \tag{4.8}$$

where CC_t is the net present value at time t of the total macroeconomic costs of the most cost-effective programme that keeps throughputs within the budget. The costs that are postponed to the future would then be

$$PB_t = b_t PC_t \tag{4.9}$$

The rest, $b_t(Z_0 - Z_t)$, is already accounted for in the national savings.

Of course the linearity assumed here is highly unrealistic. It must be taken into consideration that on the one hand postponement increases the probability of lower abatement costs due to technical progress from abroad. On the other hand postponement compresses the investment programme into a shorter period, which can lead to inflationary pressure and costly adaptation to possible price increases.

Postponed Emission Reductions as Natural Resource Consumption. The postponed costs could then replace the second term in equation (4.2):

$$D_{Nt} = (\mathbf{R}_t - \mathbf{R}_{t-1})\mathbf{q}_t + \mathbf{PC}_t\mathbf{b}_t \tag{4.10}$$

where $\mathbf{PC}_t$ is a vector of deviations from a programme of conversion to sustainable levels of emissions or uses of particular substances. This will, in most cases, be less than the total emission figures and b_t, that is the average macro avoidance cost, would in many cases be less than the average damage costs. Consequently, the total savings requirements related to emissions are likely to be less when accounted for as in (4.10) than as in (4.2).

The interpretation of D_{Nt} according to equation (4.10) will be somewhat different from that according to equation (4.2). It will be the changes of the natural capital balance as the sum of changes in assets and changes in liabilities.

In the sections on the entitlement problem, I argued that for the analysis of sustainable development of a particular economy, it is necessary also to take account of the planned growth rate of GNP per capita and the projected population growth. The net national savings required for meeting these targets can be calculated under assumptions about factor productivity growth.

These modifications and extensions of the savings rule produce a tool that is suitable for analysis of whether an economy is saving enough to accomplish the sustainability strategy defined for that economy.

5. Final Remarks

In this chapter, I have discussed the relevance of using measures of national wealth and savings to indicate or even define sustainable development. The intention has been to apply the very abstract theoretical results to the more complicated reality of country specific strategies for sustainable development.

I concluded that in most sustainability strategies it is logical to set targets in the form of a level of per capita consumption given an optimal level of environmental quality. A target such as non-declining national wealth ignores the savings requirements from industrialisation, population growth, and programmes for environmental protection and restoration. Since industrialisation population growth and environmental protection are major components of any strategy for sustainable development, it is very important to have these aspects included in a savings requirement for sustainable development. In any case, it is extremely important to be specific about the assumptions because different assumptions lead to very different results.

The assumption of resource substitutability is very basic to the whole idea of natural capital. The question discussed is not whether natural resources in are in general substitutable or not, but exactly which natural resources are substitutable and which are not. It is argued that three conditions of substitutability must be met for a given natural resource to be substitutable. In the macroeconomic terminology, it must be substitutable in the aggregate production function (the technological substitutability condition), in the wellbeing function of the representative individual (the preference ordering condition), and in the institutional interplay between the two functions (the appropriability condition). Considering each of the natural resources in turn leads to the conclusion that exhaustible and renewable resources are substitutable in their capacity of income sources while they can be so in their capacity of throughput sources. Environmental qualities, however, do in many cases, and particularly in small countries, do not meet all of the three conditions.

It all boils down to only including natural resources that contribute to GNP in the natural capital concept. This is additionally very convenient from an accounting point of view.

There is no doubt that national savings play a decisive role in sustainable in sustainable development. A savings rule can be used to make a consistency check on the whether we leave enough to the future generations to make them as able to meet their needs as we want them to be. But what are the needs of the future generations? Is it, e.g. maximum per capita consumption or just the same aggregate consumption as we enjoy?

Economic theory can provide us with the tools for analysing what it takes to reach the sustainability goals that differ from country to country. At least for developing countries, these goals include a much higher per capita GNP, eradication of poverty, and a number of preserved or restored environmental qualities. How much and how fast is an ethical and thus a political question. When we put all that together and take account of the present consumption of non-renewable resources and over-consumption of renewable resources, we can get an estimate of how much savings it really takes to achieve all the good things we want for our descendants. And it is likely to be more than what is required to keep the total capital stock constant.

References

Aage H. Economic arguments on the sufficiency of natural resources. Cambridge Journal of Economics 1984; 8: 105 - 13.

Arrow KJ, Fischer AC. Environmental preservation, uncertainty, and irreversibility. Quarterly Journal of Economics 1974; LXXXVIII (2, May): 312 – 19.

BP, The British Petroleum Company p.l.c. BP Statistical Review of World Energy 1996. http://165.121.20.76/bpstats/index.html, 1996.

Chichilnisky Graciela. An axiomatic approach to sustainable development. Social Choice and Welfare 1996; 13(2):.

Daly Herman E. Beyond Growth. Boston: Beacon Press, 1996.

Dasgupta Partha. 'Natural resources in an age of substitutability'. In Handbook of Natural Resource and Energy Economics, (vol. III) Allen V. Kneese and James L. Sweeney. Amsterdam: North-Holland, 1993.

Dasgupta Partha. 'Optimal development and the idea of net national product'. In The Economics of Sustainable Development. Ian Goldin and L. Alan Winters, eds. Cambridge: Cambridge University Press, 1995.

Dasgupta Partha and Heal Geoffrey. Economic Theory and Exhaustible Resources. Welwyn: Nisbet, 1979.

European Environment Agency. Environment in the European Union 1995. Luxembourg: Office for Official Publications of the European Communities, 1995.

Funtowicz Silvio O. and Ravetz Jerome R. 'A new scientific methodology for global environmental issues'. In Ecological Economics. The Science and Management of Sustainabilty, Robert Costanza, ed. New York: Columbia University Press.

Groot Wouter T. de. Environmental science theory. Concepts and Methods in a One-World, Problem-Oriented Paradigm. Amsterdam. : Elsevier, 1992.

Hacche Graham. The Theory of Economic Growth. An Introduction. London: Macmillan, 1979

Hamilton K. Green adjustments to GDP. Resources Policy 1994; 20 (3): 155 - 68

Hamilton K, Atkinson G. Valuing air pollution in the national accounts. Mimeo, CSERGE, University College London, 1995.

Hartwick JM. Intergenerational equity and the investing of rents from exhaustible resources. The American Economic Review 1977; 67(5): 972 – 74.

Hartwick JM. Non-renewable Resources Extraction Programs and Markets. Chur.: Harwood Academic Publishers, 1989.

Hartwick JM. Natural resources, national accounting and economic depreciation. Journal of Public Economics 1990; 43: 291 – 304.

Hartwick John M. and Olewiler Nancy D.: The economics of natural resource use. New York: Harper & Row, 1986.

Heal Geoffrey and Chichilnisky Graciela (1991): Oil and the International Economy. Oxford: Clarendon, 1991.

Krautkramer JA. Optimal growth, resource amenities and the preservation of natural environments. Review of Economic Studies 1985; LII: 153 – 70.

Malinvaud, E. Lectures on Microeconomic Theory. Amsterdam: North-Holland, 1976.

Miljø- og Energiministeriet (The Danish Ministry of Energy and the Environment). Natur- og miljøpolitisk redegørelse 1995. Copenhagen: Statens Information, 1995.

Miljø- og Energiministeriet (The Danish Ministry of Energy and the Environment). Energi 21. Regeringens Energihandlingsplan 1996. Copenhagen: Statens Information, 1995.

Mincer Jacob. Schooling, Experience, and Earnings. New York: National Bureau of Economic Research and Columbia University, 1974.

Nehru V, Dhareshwar A. New Estimates of Total Factor Productivity Growth for Developing and Industrial Countries. The World Bank, Policy Research Working Paper 1994; 1313.

Pearce, David W. Blueprint 3. Measuring Sustainable Development. London: Earthscan, 1993

Pearce DW, Atkinson GD. Capital theory and the measurement of sustainable development: an indicator of 'weak' sustainability. Ecological Economics 1993; (8): 103 – 8.

Pearce DW and Atkinson Giles D. 'Measuring sustainable development'. In The Handbook of Environmental Economics, Daniel W. Bromley, ed. Oxford: Blackwell, 1995.

Perrings Charles. 'Reserved rationality and the precautionary principle: technological change, time, and uncertainty in environmental decision making'. In Ecological Economics. The Science and Management of Sustainabilty, Robert Costanza, ed. New York: Columbia University Press.

Phelps E. The Golden rule of accumulation: A fable for growthmen. American Economic Review 1961; 51: 638-43.

Ramsey Frank P. 'A Mathematical theory of saving'. In Foundations. Essays in Philosophy, Logic, Mathematics and Economics by Ramsey FP, Mellor DH ed. London: Routledge & Kegan Paul, 1979.

Randall A. 'Total and nonuse values'. In Measuring the Demand for Environmental Quality, Braden John B and Kolstad Charles, eds. Elsevier, North-Holland, 1991.

Rawls John A. Theory of Justice. Cambridge MA: The Belknap Press of Harvard University Press, 1972.

Solow RM. Intergenerational equity and exhaustible resources. Review of Economic Studies 1974; 41: 29 – 45.

Solow RM. On the intergenerational allocation of natural resources. Scandinavian Journal of Economics 1986; 88(1): 141 – 9.

Solow RM: An Almost Practical Step Towards Sustainability.Washington: Resources for the Future, 1992.

Spash CL, Hanley N. Preferences, information and biodiversity preservation. Ecological Economics 1995; 12(3).

Tongeren Jv, Schweinfest S, Lutz E, Luna MG, Martin G. 'Integrated environmental and economic accounting: case study for Mexico'. In Toward Improved Accounting for the Environment, Lutz Ernst, ed. Washington: The World Bank, 1993.

Turner R Kerry. 'Sustainability: principles and practice'. In Sustainable Environmental Economics and Management. Principles and Practice, Turner R Kerry, ed. London: Belhaven, 1993.

Victor PA. Indicators of sustainable development: some lessons from capital theory. Ecological Economics 1991; 4: 191 – 213.

The World Bank: Monitoring Environmental Progress. A Report on Work in Progress. Washington: The World Bank, 1991.

WCED, World Commission on Environment and Development. Our Common Future. Oxford, 1987.

WRI, World Resources Institute. World Energy Projections, http://www.geocities.com/RainForest/3621/ENERGY1.HTM, 1997.

5 SUSTAINABLE DEVELOPMENT AND ECOLOGICAL MODERNISATION: *National Capacity for Rigorous Environmental Reform*

Maurie J Cohen

Oxford Centre for the Environment, Ethics, and Society, Mansfield CollegeOxford University

5.1 Introduction[1]

Attentive observers in recent years have begun to ascertain a policy shift in advanced nations as the concept of sustainable development has gained credence across a range of the political-economic spectrum. Even interests typically regarded as staunch antagonists to reform – for instance major chemical and resource extraction corporations – have begun to endorse a version of environmental sustainability. While sustainable development was once the rallying call of a few lonely environmentalists, its merit has become an accepted, and indeed almost fashionable, point of faith for many people and organisations.

One consequence of the ascendancy of sustainable development, which once referred to the radical restructuring of society in accordance with ecological limits, is that the idea no longer has any concrete meaning. Diverse parties now confer on the virtues of environmental sustainability without any shared understanding of what this concept means or how individuals would have to change their production and consumption patterns (if they have to be changed at all) to achieve such a goal. Not surprisingly, as the popularity of sustainable development has spread, nimble issue entrepreneurs have fused their own noble initiatives to its agenda. Some definitions even transcend narrow ecological objectives to encompass causes such as education reform, health-care provision, and poverty alleviation. One astute commentator has remarked that 'sustainable development is a term that everyone likes, but nobody is sure of what it means' (Daly 1996).

Observations such as these, suggesting that the aims of sustainable development are now confused and ill-defined, have become commonplace.

Nonetheless, some consensus has begun to coalesce in advanced nations around a political programme to achieve the next generation of environmental improvements (Weale 1992; Andersen 1994; Hajer 1995; Mol 1995). A vision known as ecological modernisation has provided much of the inspiration for this project. Developed initially in theoretical terms by a small group of German social scientists, ecological modernisation was first put forth as a rigorous approach for harmonising economic and social progress with environmental stewardship (Huber 1985; Jänicke 1985; Simonis 1988). As this notion has moved from academic ideal to practicable programme public officials have deflated most of its original ambitiousness and they have recast ecological modernisation as a much more modest variant known as 'eco-efficiency.'[1] This chapter begins by highlighting the relationship between these two concepts and then reviews from three different perspectives the spirited criticism they have inspired. Although there is merit to much of this critique, the following section questions the aptness of some of the indictment's harsher elements. The chapter then presents the results of an undertaking to assess comparatively the ability of a sample of advanced nations to make progress consistent with ecological modernisation. The final section formulates a few stylised hypotheses designed to generalise about the future prospects of this paradigm from the experiences of two countries purported to be in the incipient stages of this evolution.

5.2 The Theory of Ecological Modernisation

Environmental social scientists formulated the theory of ecological modernisation during the 1980s as accumulating evidence began to suggest that the regulatory approaches put into place during the previous decade were inadequate (Commoner 1990; Weale 1992). Rather than address the root causes of environmental degradation, this generation of environmental policies, with their emphasis on end-of pipe technologies and bureaucratic oversight, tended to displace pollutants across both time and space. Moreover, prescriptive strategies presumed strong causality between the perpetrators and victims of ecological harm that proved difficult to sustain in practice. Experience confronting such dilemmas as acid rain, toxic contamination, and global warming has highlighted

[1] A number of related terms, such as environmental excellence, eco-excellence, strategic environmental management, and enhanced resource productivity have also begun to come into usage. In the United States, some authors have advanced the expression 'third wave environmentalism' (see, in particular, Dowie 1995).

the ambiguity of these connections and underscored the difficulty of achieving precise empirical evaluation.

Ecological modernisation offers a means for transcending the current impasse to achieve environmental improvement by promoting a fundamental shift in human interactions with the biosphere. The essential feature of this new paradigm is a technological restructuring by which cleaner, more efficient, and less resource-intensive production processes will launch advanced nations to the next stage of social development. The redesign of existing institutions to enhance their capabilities for protecting human health and the natural environment will correct those industrial systems that have contributed to environmental degradation. This new phase of development will surmount the conventional zero-sum relationship between progress and the preservation of ecological integrity by making it possible to delink economic activity from its environmental costs. In other words, this construct reframes the terms of discourse by interpreting pollution reduction as a means of enhancing economic competitiveness rather than as an externality requiring the installation and maintenance of expensive remedial technologies.

Though there is undoubtedly some variation among the different authors who have contributed to the theory of ecological modernisation, this paradigm can be expressed in terms of six overarching principles.

First, ecological modernisation envisions a process of *superindustrialisation* to correct the so-called design flaws of modernity. This new industrial revolution entails the design and implementation of ecologically efficient manufacturing processes that will override the need for corrective treatment of pollutants and other harmful byproducts. It is through the advent of these new techniques that advanced nations can sever the coupling between economic growth and biospheric deterioration, thus propelling themselves onto a new developmental trajectory.

Second, rigorous environmental reform requires the implementation of *strict government regulations* to promote the innovation required for industrial transformation. While ecological modernists do not dismiss the use of voluntary agreements out of hand, they emphasise the use of increasingly tighter regulatory controls as necessary for forcing required technological changes. Through this mechanism domestic producers will be able to capture and retain first-mover advantages in export markets for environmental technology and 'green' products, a key feature for the initiation of positive-feedback cycles.

Third, prevailing policy in most advanced nations still encourages trans-media movement of contaminants, for instance from air to land, and this approach simply moves problems from one domain to another. Ecological modernisation attempts to overcome this dilemma by shifting emphasis in favour of *integrated pollution management.* This component of the paradigm encourages the use of systems theory to redesign both manufacturing processes and regulatory procedures to correct the reductionism that contributes to conventional single-media interventions.

Fourth, in most advanced nations the high cost of implementing risk-reduction measures and the prevalence of scientific uncertainty are often used as justifications to postpone action to mitigate hazards to human health and the environment. Ecological modernisation confronts these delays by appealing to the use of *anticipatory planning practices.* The German notion of *Vorsorgeprinzip*, or the precautionary principle, provides the prototype. This concept contends that the lack of scientific certitude is insufficient reason to postpone the taking of prudent measures to alleviate potential harm. The precautionary principle recognises that scientific research will likely generate additional evidence over time and that public officials should make periodic policy adjustments as this new information becomes available.

Fifth, ecological modernists endorse the *organisational internalisation of environmental responsibility* consistent with the Dutch principle of *verinnerlijking*. This concept requires all public and private entities to integrate a concern for environmental quality into all of their activities as a means of overcoming the standard approach of treating ecological factors as add-on considerations. Furthermore, ecological modernisation is inconsistent with the conventional approach in which governments and corporations delegate environmental concerns to a specific organisational entity (e.g., Environment Ministry) with the expectation that this arm of the bureaucracy will ensure responsible stewardship. This paradigm asserts that an environmental commitment must become an intrinsic element of every decision and this philosophy must pervade all levels of organisation

Finally, antagonism and conflict have come to characterise the environmental policy practises of many advanced nations. Political scientists normally identify the United States as the consummate example of this policy style, but contentiousness has infiltrated other national contexts to varying extents. Ecological modernists claim that it is necessary to move beyond divisiveness and to encourage the building of *constructive relationships* among industry, government, and the public grounded in good faith and the free exchange of information.

As ecological modernisation has begun to move during the past decade from the realm of theory to the world of practical application, specific portions of the concept have received guarded endorsement. Certain multinational corporations have focused on the prospects ecological modernisation affords for technological innovation. In some countries, a small number of exceptionally strategic firms have begun to encourage government to adopt stricter environmental controls as a means of forcing this turn. Furthermore, several national governments, as well as leading international organisations such as the European Union and the Organisation for Economic Cooperation and Development (OECD), have endorsed portions of this reform programme (Hajer 1995; Gouldson and Murphy 1996; OECD 1996a). Despite this progress, the subtler dimensions of ecological modernisation – such as the precautionary principle and the internalisation of environmental responsibility – have met resistance, even in these otherwise favourable contexts. The lack of headway on these constituent elements can be attributed to two factors. First, these principles can, if applied indiscriminately, compromise short term economic growth by forcing the abandonment of undertakings that may be ecologically harmful. Second, managers cannot easily instill a precautionary sensibility or an organisational philosophy that encourages integrated decision making. Rather, these adaptations require individuals and organisations to formulate a new ethic that treats the natural environment as more than a resource for human exploitation. It is in the effort to come to terms with these principles that advanced nations will confront the true challenges of sustainability. Due to their hesitancy to face these difficult choices, a process of selective adoption has precipitated a metamorphosis in which business managers and policy officials have replaced ecological modernisation with the less challenging concept of 'eco-efficiency.' This alternative programme does not embrace a call for large-scale societal transformation, but rather encourages the design of prescriptive interventions to promote environmental improvement that are largely consistent with customary procedures.

Eco-efficiency refers to the implementation of industrial practices that can improve corporate profitability by redesigning manufacturing processes to reduce the production of waste at the source to avoid remedial treatment (Makower 1993; Fischer and Schot 1993; Welford 1996). Essentially an extension of the quality revolution of the 1980s inspired by Edwards Deming and Joseph Juran, eco-efficiency encourages firms to seek 'win-win' solutions that promote simultaneously economic profitability and environmental improvement (see also Cairncross 1992). Numerous case studies, principally of large industrial corporations, have confirmed the availability of such gains and commercial interest has spawned a coterie of consultants to assist firms capitalise on these opportunities. The new professional field of industrial ecology is dedicated to pressing these achievements beyond narrow managerial modifications (e.g., shop-floor reorganisation, environmental auditing) and aims to develop a range of ambitious innovations including closed-loop manufacturing systems and zero-emission vehicles (see, for example, Ayres 1996; Socolow 1997).

Eco-efficiency first received popular attention in 1990 when Michael Porter, a leading management strategist at the Harvard Business School, gave his endorsement. Specifically, Porter contended that environmental improvement offered valuable opportunities for firms to enhance their competitiveness (Porter 1990, 1991; see also Porter and van der Linde 1995a, 1995b). The cause was then joined during the lead-up to the Earth Summit in Río by the World Business Council for Sustainable Development (WBCSD), an international consortium of major industrial corporations that includes Dow Chemical Company, Asea Brown Boveri, and Volkswagen (Schmidheiny 1992a, 1992b; Schmidheiny and Zorraquín 1996). In the interim, eco-efficiency has become the consensus strategy among multinational corporations more generally for striving toward some modicum of environmental sustainability. Governments have also embraced the concept, hopeful that this approach will provide a means for transcending more politically precarious interventions. Even once-radical pressure groups such as Greenpeace have begun organising seminars to advise business managers on how to proceed down this path (*The Economist* 1995).

5.3 Critiques of Ecological Modernisation/Eco-Efficiency

Despite these apparent accomplishments, several groups of scholars remain dismissive of ecological modernisation/eco-efficiency and their critiques have proceeded from three different perspectives: environmental economics, social theory, and policy analysis. Environmental economists have focused on relatively narrow technical issues and, as such, their criticism applies only to eco-efficiency. Social theorists and policy analysts have set their sights higher and have raised questions about the soundness of the entire ecological-modernisation paradigm. The following section examines each of these critical reviews.

5.3.1 Environmental Economics Critique

Environmental economists have confined their criticism to eco-efficiency and have not treated this approach as a constituent element of a larger reform programme (Palmer *et al.* 1995; Portney 1995; see also Walley and Whitehead 1994). Their principal cause for wariness rests on an evaluation that eco-efficiency violates standard economic assumptions about competition and underestimates the true costs of environmental protection. While the potential for eco-efficient gains is surely attractive to contemplate, such free lunches (or indeed paid lunches) are rarely realisable in practice because firms normally operate near their efficiency frontiers. A small group of American economists, who discount the likelihood of finding $10 notes on the ground, has been primarily responsible for developing this line of critique. Such fortuitous events are exceptional because, as we all know, someone else would have already picked up the prize in most instances. In other words, firms conduct business in a competitive milieu and managers do not need the assistance of environmental regulators to guide them toward profitable solutions. If such opportunities existed, scrupulous managers would have seized upon them. To Paul Portney's mind the whole notion of win-win solutions is preposterous. To graphically illustrate his point he portrays proponents of eco-efficiency as taking the position of 'I'm here from the US EPA [Environmental Protection Agency] and I'm here to help you maximize shareholder value' (Portney 1995).

From this point of view, economists' dismissal of eco-efficiency proceeds on two fronts. First, claims alleging the reconciliation of profit maximisation and environmental improvement are illusory. The case studies conducted to demonstrate the existence of such solutions, they argue, are inconclusive and are not representative of economic conditions more generally. While certain innovation offsets may compensate for initial outlays to achieve regulatory compliance, experience indicates that gratis effects will in the large majority of cases be inconsequential in comparison with costs. Even in instances where such opportunities persist, it is unlikely that regulators will be able to guide managers to these solutions through the use of strategic environmental policy. Furthermore, the allocation of scarce financial resources and personnel to the control of environmental pollution ultimately displaces investment from more 'productive' uses, thus dampening profitability and reducing economic growth.

Second, certain unique circumstances may exist in which strong environmental regulations will enhance firms' international competitiveness, but it is just as plausible, according to economic theory, that weak controls will achieve the same result. In particular, lax regulations enable domestic producers to drive down costs relative to their global competitors. It is also instructive to recognise that for most firms the expense of complying with environmental regulations is a small portion of their total outlays. Much larger differences in other costs of production (e.g., labour, materials) will overwhelm any competitiveness-enhancing gains they may realise in this sphere.

In summary, while environmental economists concede that eco-efficiency may be able to achieve certain improvements at the margin, implemented with more expansive expectations, it is a strategy fraught with false promises. Unfortunately, environmental protection is an expensive endeavour and money-saving short-cuts are unlikely to ease this burden. More appropriate would be to continue to use standard economic tools to compare the social benefits of prospective regulatory measures with their private costs and to aim for more effective expenditures of resources.

5.3.2 Social Theoretical Critique

The social theoretical critique is unabashedly more expansive than that advanced by environmental economists and takes on the grander reform programme contained under the rubric of ecological modernisation. Social theorists are critical of this paradigm because it fails to address basic distortions and inequities of industrial capitalism. David Harvey (1996) has emerged as the leading expositor from this perspective and it is worthwhile to focus in some detail on his concerns (see also Sachs 1993; Hajer 1995, 1996; Christoff 1996).

Harvey considers ecological modernisation (and by implication eco-efficiency) to be little more than a modern-day rendition of the management principles that Gifford Pinchot and other so-called utilitarian environmentalists popularised in the United States at the turn of the century.[2] Although he does not discount entirely the potential of this approach to deliver certain, delimited forms of improvement, he is quite pessimistic of its likelihood for achieving anything approaching a genuine form of sustainable development. For Harvey, ecological modernisation is little more than a ruse to evacuate the radical content from political discourse concerning sustainability and to use the current phase of public environmental interest to rationalise currently fragmented and *ad hoc* ecological management systems.

The social theoretical critique emphasises that ecological modernisation is expressed principally in the language of science. Economics, chemical engineering, and toxicology are idioms that powerful organisations can understand and manipulate to their advantage. By way of contrast, the passion, morality, and sentimentality that often accompany political mobilisation around ecological issues is subject to dangerous unpredictability and are difficult discourses to use as tools for ensuring social control. For corporate managers and public officials, environmental commitments unrestrained by scientific reason endanger economic well-being. Such recklessness is endemic in the expensive, problem-plagued cleanup of hazardous-waste sites in the United States and the imposition of prohibitions on biotechnology research in Germany.

---

[2] On this point, refer also to Hajer (1996), and Hays (1959). The concepts behind this movement actually have their origins in the work of George Perkins Marsh. See, in particular, his book *Man and Nature* originally published 1864.

The ascendancy of ecological modernisation is due to the need for elite groups to assert their influence in the debate on sustainable development.

Harvey pursues this line of critique further, identifying several penetrating factors to explain why the contemporary environmental debate has crystallised around ecological modernisation and its variant eco-efficiency. It is instructive to focus on (and in some instances extend) three of his observations.

First, despite some ephemeral increases in growth, the ground was collapsing under the feet of mainstream environmental pressure groups during the 1980s as conservative governments in several advanced nations – most notably the United States and Britain – systematically began to dismantle the supports undergirding the welfare state. While this political climate proved favourable for scientifically-guided interventions based on market mechanisms, it disabled efforts to tighten controls by mobilising mass publics in support of legislative initiatives. These developments reinforced tactical missteps by several of the major pressure groups and few activist organisations developed the expertise necessary to operate effectively under the new rules (see also Dowie 1995).[3] Weakened in the extreme and shorn of their radical colouring, these organisations have discovered that by adopting the reasonable tone of ecological modernisation they can reaffirm their relevance. Though the environmental movement still retains a radical fringe in most advanced nations, these particular groups are beset by numerous internecine problems and their decidedly anti-modernist commitments are never likely to be more than a curiosity for most segments of the public.

Second, Harvey contends that concern for natural limits, future generations, and environmental sustainability provides governments of the world's most affluent countries with an effective means of diverting attention from politically contentious issues of global equity. By focusing international discussions on population growth (essentially an issue for developing countries), industrialised nations are able to deflect calls to reduce their own environmentally-destructive lifestyles. As he observes, 'The supposed sheer physical inability of the planet to support global populations with aspirations to a living standard of Sweden or

[3] Partial exceptions to this characterisation include the Natural Resources Defense Council (NRDC), the Environmental Defense Fund (EDF), and Resources for the Future (RFF) in the United States. These entities have strong scientific capabilities and are not mail-order membership organisations in the conventional sense. They have all gained political influence in recent years at the expense of mainstream groups such as the Sierra Club. Similarly, in Britain popular environmental organisations such as Greenpeace and Friends of the Earth have appreciably upgraded their scientific expertise. This new professionalism was ably demonstrated during the 1995 *Brent Spar* episode, a controversy over the disposal of a Shell-owned North Sea oil platform.

Switzerland became an important political argument against even trying to spread the benefits of capitalist growth around.'

Finally, the uniform opposition of economic interests to environmental improvement finally broke during the later portion of the 1980s as several multinational corporations began to recognise the profit potential in stronger national regulations and 'green' technologies. Support for these initiatives was predicated on a recognition that more stringent legal controls could provide useful barriers to entry, preserving lucrative markets for the most technically competent firms. Additional impetus for more responsible environmental performance has come from new activist groups that have successfully opened new ground for reform through consumer, employee, and shareholder channels (Cohen, 1998a). This new configuration, marked most evidently by the defection of firms such as 3M and Monsanto from the camp of resolute defenders of the environmental status quo, exposed many of the familiar objections put forward to stymie proponents of new ecological measures.

In summary, Harvey contends that ecological modernisation is nothing more than a discursive technique that 'internalises conflict.' He dismisses neither the concept's 'radical populist edge' nor the fact that it devotes 'serious attention to environmental-ecological issues.' Harvey, however, takes ecological modernisation to task for addressing 'the accumulation of scientific evidence of environmental impacts on human populations without challenging the capitalist economic system head on.'

5.3.3 Policy Analysis Critique

Much more disparate in its articulation to date, the policy analysis critique can be distilled into two perspectives. The first strand maintains that many of the gains ascribed to ecological modernisation are not necessarily due to successful efforts to redesign manufacturing processes. Rather, many of these improvements are attributable to the relocation of pollution-intensive activities to distant locations. In this sense, the affluent nations are simply shifting their industrial hazards to the developing world (Shrivastava 1995). While lower labour costs have provided much of the motivation for this migration, the process has promoted the relocation of industrial activities to jurisdictions with less stringent environmental controls. Even those regulations that do exist in developing countries are typically subject to lax enforcement due to inadequate expertise and institutional limitations.

Critics arguing from this perspective assert that though domestic gains in the advanced nations may contribute to a patina of ecological reform, in actuality they only exacerbate global inequities. As a cost of their late industrialisation, ill-equipped developing countries are now encountering new environmental hazards producing export goods for consumers in affluent countries. In the case of international ecological problems such as global warming due to the emission of carbon dioxide and other greenhouse gases, domestic gains arising from industrial migration are nullified (and often overwhelmed) when production commences elsewhere.

The second thread in the policy analysis critique contends that ecological modernisation does little (and eco-efficiency does even less) to address the politically vexing issue of unchecked household consumption (During 1992; Redclift 1996; Myers 1997). Despite more than two decades of increasing ecological consciousness, most people in the wealthy nations continue to evidence a firm disinclination to modify their lifestyles. They prefer instead to call for superficial adjustments at home while castigating developing nations to reduce population growth. While gains through more intensive use of raw materials and recycled supplies will certainly aid efforts to promote environmental sustainability, adherents of this perspective maintain that it is imprudent to expect sufficient material efficiencies through technological ingenuity alone (Daly 1996).

Governments have strongly resisted suggestions that environmental sustainability will require adjustments in quality of life (as measured in material terms). Even purportedly progressive leaders such as American vice-president Al Gore and ex-Norwegian Prime Minister Gro Harlem Brundtland have not taken this step (Gore 1992; Brundtland 1994; see also World Commission on Environment and Development 1987). Nonetheless, critics of ecological modernisation assert that constraints on consumption will be central to attempts to situate global society within its ecological limits and reform programmes that side-step this issue in the interests of political viability will come up short.

5.4 Ecological Modernisation Confronts its Critics

Taken in combination, these three critiques seem to present a powerful indictment of ecological modernisation. Adversaries of this paradigm range from the environmental economists who claim that it is built on unrealistic premises to social theorists who imply ecological modernisation offers an insufficiently radical programme for social change. Covering some of the ground between these two positions, we find policy analysts of different stripes who are sceptical of both the equity implications and the technological prowess of ecological modernisation. Furthermore, these critics doubt the potential of a reform programme that fails to take effective action on reducing household consumption.

In mobilising a response to defend ecological modernisation against these allegations it is useful to confront first the concerns of the environmental economists. This critique actually reflects the remoteness of most economists to real managerial decision making and the implausible assumptions underpinning this perspective derive largely from this unfamiliarity. Conceiving of firms as operating at or near their efficiency frontiers is a view that stands in stark contrast to the experience of most managers. Firms do not operate in the economists' world of perfect information and incessant competition; rather they engage in a daily struggle to assess and integrate contradictory information. Channels of communication are often faulty and demands on managers' time and energy typically exceed human and organisational capacities. Under such circumstances, cost-saving interventions frequently go unnoticed or fail to achieve sufficient support to engender action. The extensive number of case studies accumulated to date provides compelling evidence that, contrary to the economists' cautionary counsel, large numbers of 'win-win' solutions are readily available for managers inclined to look for them (see, for example, Fischer and Schot 1993; Weizsäcker *et al.* 1997).

Moving to the social theoretical perspective, one invariably finds much merit to this critique. Social theorists' hesitancy to embrace ecological modernisation as an ambitious programme for social transformation are well founded. For all its revolutionary quality, ecological modernisation offers a prescription that, for the most part, goes with the grain of the time. Such an approach is unlikely to correct the injustices that plague modern capitalism and will not promote a version of sustainable development that conforms with most ardent environmentalists' ideological commitments. Moreover, social theorists' tacit assertion that ecological modernisation conveys an enthusiastic technological determinism deserves careful evaluation. Important to bear in mind is the fact that while this paradigm places considerable emphasis on engineering solutions to environmental dilemmas, it is not motivated by a sense of triumphalism. Ecological modernisation offers a relatively modest vision of social transformation, one exclusively focused on tempering individual and organisational behaviour to promote environmental stewardship. Critics should evaluate this approach on the basis of its pragmatism rather than its failure to advance a more ardent agenda.

More difficult is the task of defending ecological modernisation against the policy analysis critique. These critics claim, in part, that this paradigm is little more than a shell game that moves the environmental burden of industrial production onto the backs of the world's most vulnerable people. An extensive body of literature has accumulated indicating that globalisation is encouraging multinational corporations to migrate to havens where they can operate with minimal public oversight (for a review, see Pearson 1987). In particular, countries such as Mexico, India, and China provide attractive relocation destinations for pollution-intensive activities, though the data is mixed on the extent to which environmental issues are inducing this movement. Regardless of the actual motivation, the evidence suggesting that the new geographic distribution of production is shifting environmental hazards from affluent to developing countries remains quite compelling. Empirical analyses to disentangle displacement effects in advanced nations from more meaningful gains due to technology enhancement poses very complex methodological problems.

The second strand of the policy analysis critique addresses the failure of ecological modernisation to address the demand side of the environmental equation. This is a trenchant point and it remains questionable whether the fruit of human ingenuity, in the form of technological advances, will come fast enough to stem an expanding pattern of global consumption arising from both population growth and increasing affluence. Normally, consumption is expressed narrowly in terms of the depletion of non-renewable resources. However, this issue is more appropriately addressed in terms of the spread of consumerism. Without some concerted effort to reduce the overall volume of material throughput, especially in the advanced nations, it becomes highly questionable whether technological improvements will enable humanity to make the leap in societal organisation envisioned by ecological modernisation.

Consideration of the efficacy of ecological modernisation has not been proceeding in the measured terms portrayed in the preceding discussion. In actual practice, selective adaptation of particular principles is leading to a process of partial and incomplete diffusion of this paradigm. Ecological modernisation has had some success as a policy discourse because it provides economically and politically influential institutions with an effective rhetorical technique for redefining the sustainable development agenda. This reframing simultaneously delegitimises steadfast opposition and defuses appeals for more extreme restructuring that might compromise the hegemony of elite organisations.

Under these circumstances, are we then to conclude that ecological modernisation is incapable of achieving little more than modest engineered improvements of manufacturing processes that do not impinge on existing prerogatives? Does ecological modernisation simply exceed the current capacity and inclinations of advanced nations for social transformation? These formidable questions are difficult to answer definitively at this stage, largely because of the desultory way in which ecological modernisation has so far proceeded in actual practice. Changes consistent with this theory may indeed be occurring; however, their rudimentary and incomplete qualities make it difficult to ascertain their progress with any certitude. To get a handle on ecological modernisation in specific national contexts, it is necessary to construct a tangible marker capable of measuring the relationship between environmental quality and economic activity. We must also recognise that an overall notion of environmental quality embodies numerous constituent elements, for example air and water quality, natural resource management, transport mobility, biodiversity. In an effort to delimit the scope of analysis, the following study focuses on the emission of pollutants to air as a proxy for developments in environmental policy more generally.

5.5 Ecological Modernisation in National Contexts

For ecological modernisation to become a practicable consideration a country must have attained a requisite level of economic and social advancement (Cohen 1997; Jänicke and Weidner 1997). Accordingly, the pool of suitable nations for this study effectively consists of the top-tier members of the Organisation for Economic Cooperation and Development (OECD): Austria, Belgium, Canada, Denmark, Finland, France, Germany, Italy, Japan, the Netherlands, Norway, Sweden, Switzerland, the United Kingdom, and the United States.

Air quality has long been a focus for policy attention and was a key issue during the first wave of environmental reform during the 1970s, largely because it represented one of the more odious forms of pollution. In recent years, due to anxiety about the contribution of carbon-dioxide emissions to global warming, we have again become mindful of the effects of current production and consumption patterns on the atmosphere. Regulatory interventions in the advanced nations first called for the use of tall smokestacks to facilitate dispersal over wide geographic areas. When the international community discovered, largely due to the reproaches of the Scandinavian countries, that this strategy contributed to acid deposition, flue-gas desulphurisation equipment (scrubbers) became the preferred option. In more recent years, owing to complications associated with the disposal of sulphurised sludge, the current vogue is for the redesign of emission sources. At the same time the advanced nations were pursuing this series of options, many of them were also giving strong endorsement to nuclear power – most prominently in Sweden and France – as a means of reducing air pollution from electricity production. Elsewhere government initiatives encouraged the substitution of low-sulphur fuels (e.g., natural gas) for oil and coal. The tenacity of the dilemma concerning air quality and the numerous alternatives that the public and private sectors have experimented with over the past three decades suggest there has been considerable opportunity for social learning in this sphere. Accordingly, it is in this policy area that we might reasonably expect to glean the emergent signs of ecological modernisation.

As indicated above, in assessing progress moving toward ecological modernisation across several nations it is necessary to devise a criterion that can signal the manifestation of changes consonant with this paradigm. Normally, we would expect the production of pollutants released into the atmosphere (as measured by volume) to be a function of overall economic output and for time series of these two measures to move roughly in parallel with one another (refer to Figure 5.1A). Countries taking their first steps toward ecological modernisation should be evidencing some indication of a developmental pattern that breaks the connection between economic growth and pollution generation. This alternative relationship, in which the two time series become increasingly delinked and veer off in opposing directions, is depicted in Figure 5.1B.

With these preliminary issues resolved, we can examine some data on selected releases to air for each of the fifteen candidate countries. In each case, five separate emissions streams are considered: sulphur oxides (SO_x), nitrogen oxides (NO_x), particulate matter, carbon monoxide (CO), volatile organic compounds (VOCs), and carbon dioxide (CO_2). Indexed values were calculated (1980 = 100) for each type of emission, as well as for gross domestic product (GDP).[4] These air pollutant values are output indicators of environmental quality and it is useful to contrast this methodology with complimentary approaches that examine the environmental effects of temporal changes in economic structure and input measures of material throughput (see, for example, Jänicke *et al.* 1989).

---

[4] Emissions indexes are calculated from data in the OECD's *Environmental Data Compendium* 1995. Because of different methodologies used to measure emissions for each nation, the comparison of trends across countries should be pursued with care. This difficulty is not especially germane to the current analysis because the data are treated as discrete time series and no inter-country comparisons are undertaken. GDP data are from the IMF's *International Financial Statistics Yearbook* 1996.

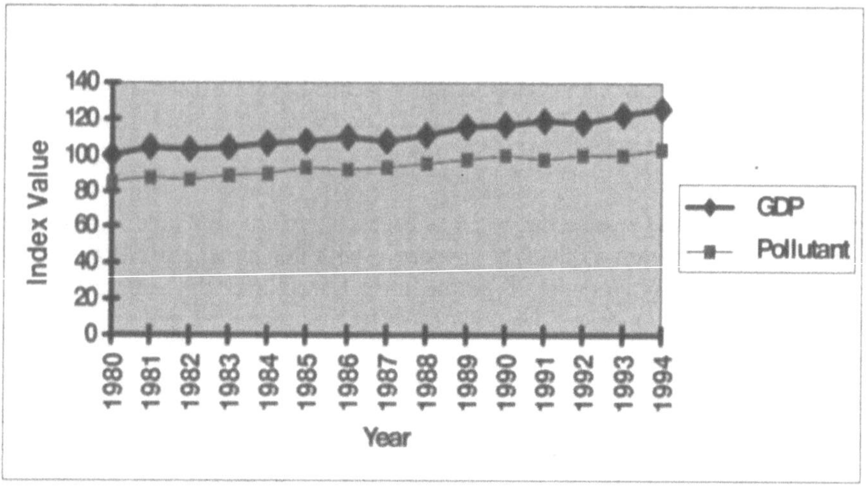

Figure 5.1A Conventional pollution-economic growth pattern

These data suggest that Germany is beginning to decouple economic growth from its SO_x emissions in a manner consistent with ecological modernisation (refer to Figure 5.2A). Though several advanced nations achieved sharp improvements in SO_x efficiency during the 1980s due to concern about acid rain, Germany is unique for the rapidity of its gains. The country has curtailed SO_x production by substituting low sulphur coal, as well as through technological means. In fact, German manufacturers' ability to design state-of-the-art scrubbers and to sell this equipment internationally has become an exemplary case of how strict environmental regulations can help in the cultivation of lucrative export niches (Moore and Miller 1994). More generally, Germany's recent development pattern supports the conclusion that a discernible delinkage between economic growth and resource utilisation is occurring (see also OECD 1993). Such accomplishments, in combination with collateral evidence suggesting a growing commitment to ecological responsibility within the German business community and the establishment of a supportive public policy framework, indicate meaningful movement toward ecological modernisation. Compromising the country's environmental leadership, however, have been some especially entrenched special interests – notably in transport and agriculture – that have managed to stifle more vigorous action. Also threatening to slow down the wheels of environmental progress has been the tremendous cost of German unification, as well as the country's more recent period of economic stagnation (Jänicke and Weidner 1997). Such developments have placed the country's venerated ability to foster political consensus during the post-World War II era under serious strain.

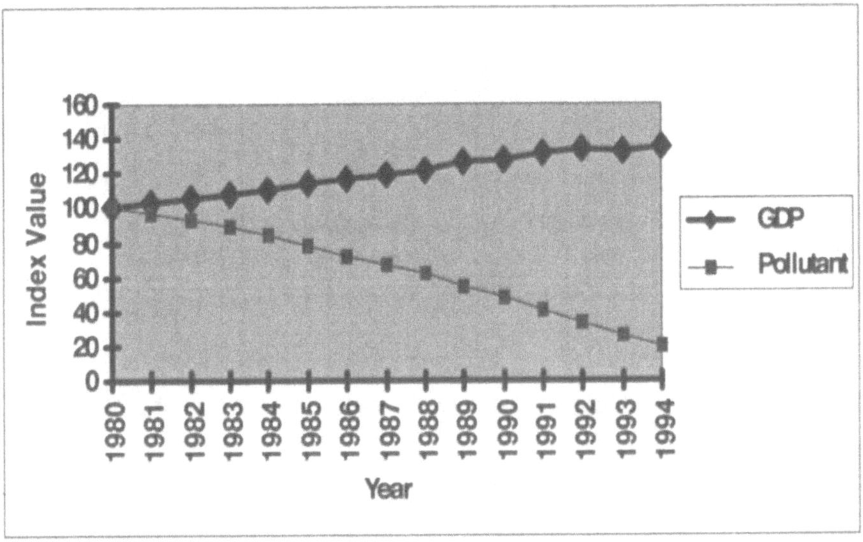

Figure 5.1B Ecologically modern pollution-economic growth pattern

More qualified assessment points to the Netherlands as a potential aspirant to ecological modernisation (Figure 5.2B). The Dutch have had considerable success increasing GDP per ton of SO_x while continuing to experience healthy economic growth throughout much of the period since 1980. An important factor contributing to this improvement is enhanced efficiency in the oil-refining sector where SO_x production was reduced by 35 percent between 1980 - 1989

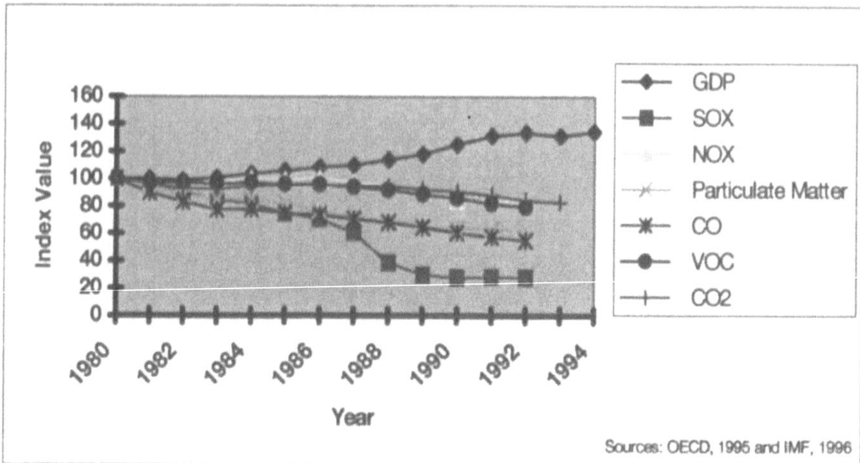

Figure 5.2A Indexed GDP and Selected Emissions to Air Germany, 1980 - 1994 (1980 = 100)

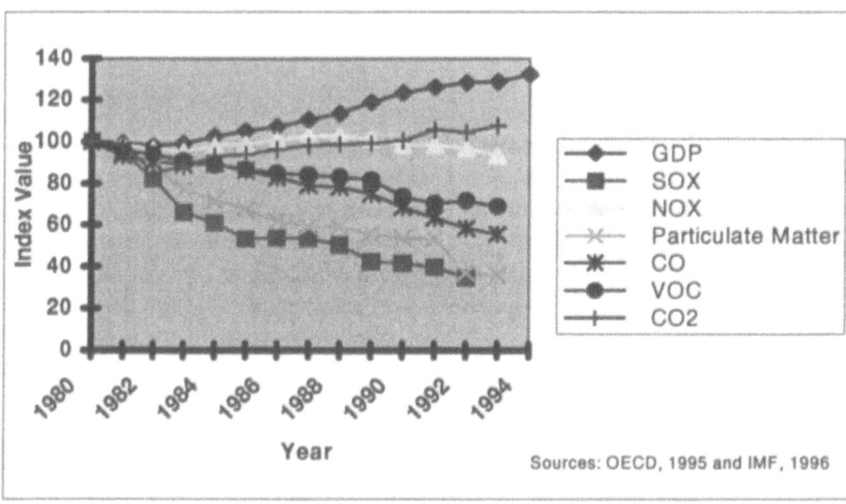

Figure 5.2B Indexed GDP and Selected Emissions to Air The Netherlands, 1980 – 1994 (1980 = 100)

(Opschoor and van der Straaten 1993; OECD 1995). The Netherlands has also moved forward in intractable areas such as particulate matter, VOCs, and carbon monoxide. Standing in contrast to these gains, however, the Dutch have failed to delink their CO_2 emissions from economic growth. Additionally, the government's willingness to grant special dispensation to politically influential corporations has regularly stymied more aggressive reform (see, for example, Opschoor and van der Straaten 1993; Hajer 1995, pp. 244-251). The Netherlands has also displayed a tendency to promote initiatives that reinforce the country's image as an international environmental leader without providing sufficient resources to ensure the achievement of concrete results (Bressers and Plettenburg 1997). These observations suggest that the Dutch are progressing toward ecological modernisation with an ambivalent commitment, one that is simultaneously purposeful and apprehensive.

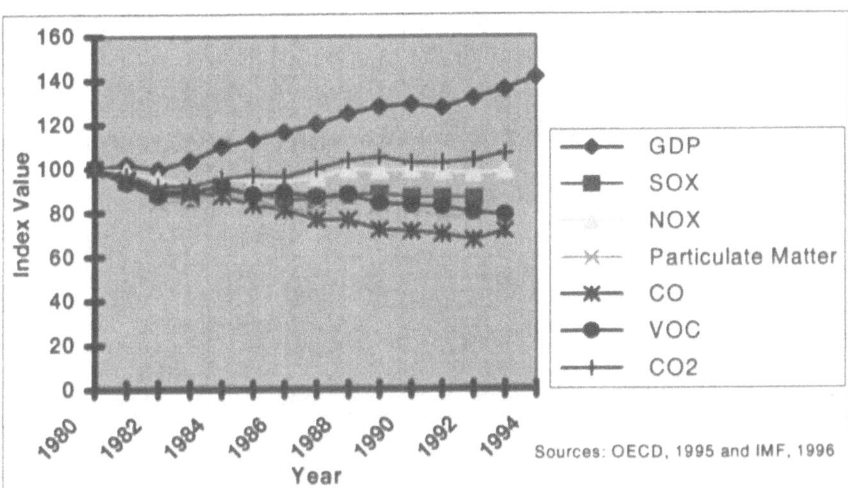

Figure 5.2C Indexed GDP and Selected Emissions to Air United States, 1980 – 1994 (1980 = 100)

The United States provides an interesting and complex case that deserves more detailed examination than is possible here. In brief, the country appears, in the face of robust economic growth during the past decade, to have stabilised its production of all measured air emissions roughly at 1980 levels. The US has not realised the SO_x efficiency gains achieved in Germany and the Netherlands during the previous decade, but has managed to sustain considerable economic growth without commensurate increases in pollution emissions to air (refer to

Figure 5.2C). Despite apparent improvements in resource productivity, US pollution emission levels remain, when compared to its OECD partners, enormous on the basis of both absolute and per-capita volumes. Nonetheless, this initial evidence suggests that the US may be embarking on a developmental trajectory in which the country attempts to 'grow out of its environmental problems.' Tempering this observation is the possibility that the country has already realised most of its relatively painless improvements. Further environmental reform to bring emission efficiencies into line with northern European levels will likely require more fundamental restructuring (e.g., ecological taxation, transport reform) and there appears to be insufficient political will in the country to pursue such exacting measures (Andrews 1997). The avowedly litigious system for dispute resolution in the US compounds this difficulty. Continued encouragement of such antagonism will make it extremely arduous to secure the cooperative solutions on which the next phase of environmental improvement will depend (see also OECD 1996b, p. 35). Under these indeterminate circumstances, the US is provisionally classified as a case of 'weak' ecological modernisation and marked as worthy of more careful scrutiny over the course of the next several years.

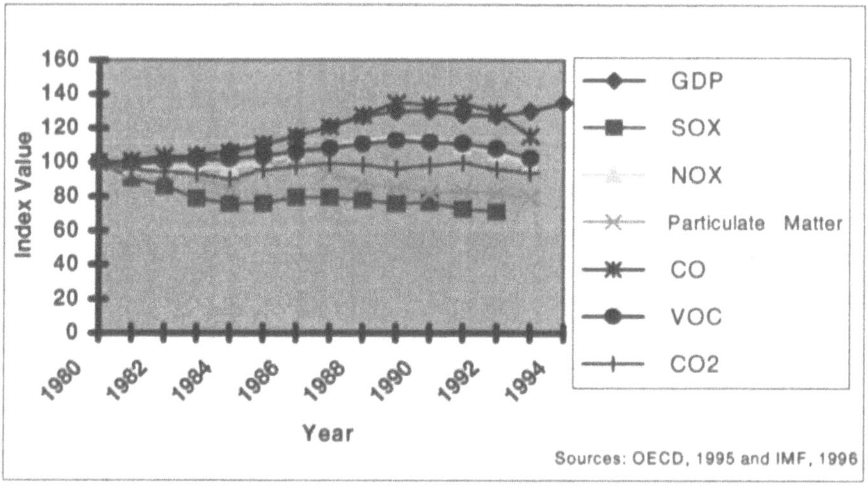

Figure 5.3A Indexed GDP and selected emissions to Air United Kingdom, 1980 – 1994 (1980 = 100)

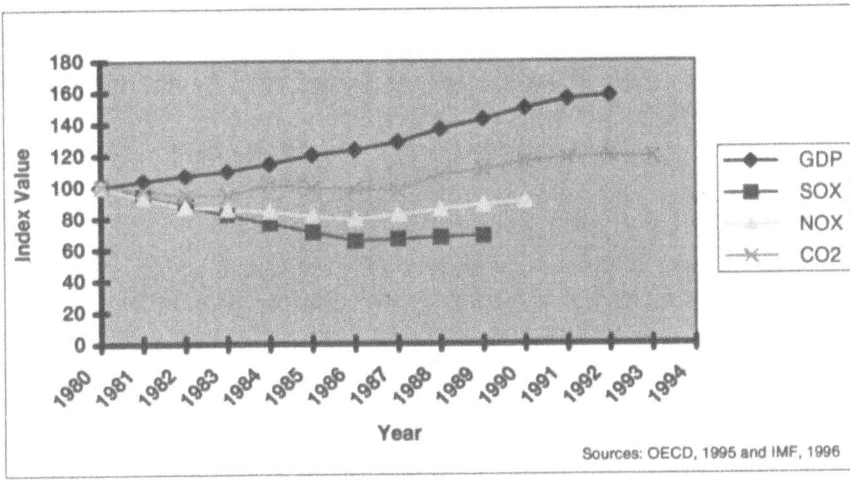

Figure 5.3B Indexed GDP and selected emissions to Air Japan, 1980 - 1994 (1980 = 100)

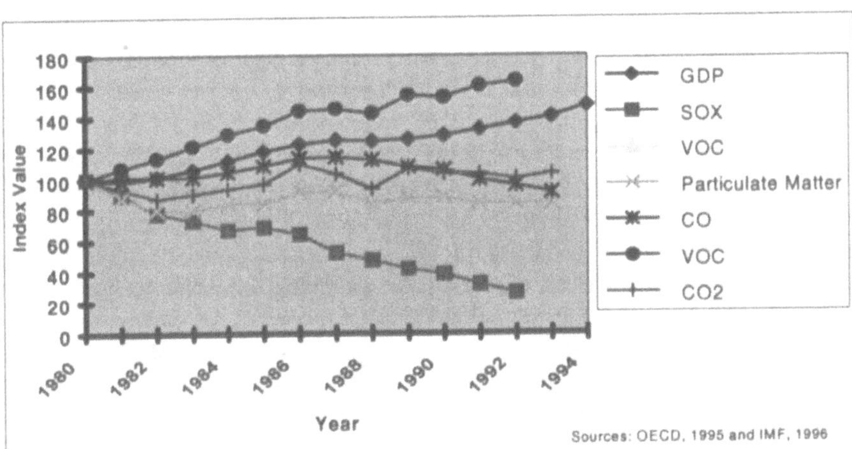

Figure 5.3C Indexed GDP and selected emissions to Air Norway, 1980 – 1994 (180 = 100)

The data on emissions to air indicate that the other candidate countries have not yet begun to disengage their economies from the environment. Such evidence implies that there has been little quantifiable progress toward ecological modernisation in these nations and their modern economic structures remain intact. Data for the United Kingdom, Norway, and Japan are presented in Figures

5.3A, 5.3B, and 5.3C respectively as representative examples of the countries in this category. The status of Sweden and Switzerland remains uncertain, in part due to insufficient data.

5.6 Conclusion

An attempt to generalise about the prospects for ecological modernisation from the handful of examples outlined above raises numerous imponderables. Cross-national differences in institutional capability, economic structure, and political culture confound efforts to extrapolate from the particular experiences of individual countries. Despite these complexities, the preceding assessment evokes an irrepressible urge to compare national potentialities for ecological modernisation and to identify exemplars worthy of emulation. It is true that ecological modernisation may not promote the radical social restructuring deemed desirable by some of its critics. Moreover, the theory, as we have seen, is deficient on several grounds: most notably its tendency to encourage (at least implicitly) the exportation of environmental hazards to newly industrialising countries and its failure to appreciate the need to temper consumption in the wealthy nations. Noting these caveats, in its stronger expressions this paradigm *does* provide a practicable blueprint for charting a course toward a more sustainable future.

It is in the spirit of these observations that this concluding discussion engages in some speculation about the societal features that have enabled Germany and the Netherlands to secure positions as inchoate ecological modernisers. To be certain, neither of these countries provides an unambiguous example of this paradigm. They are, however, the best cases currently available for testing our theoretical insights against empirical evidence. Proceeding in such a way is not meant to suggest that all advanced nations will follow a similar Rostowian road to the next phase of modernity (see Rostow 1960). On the contrary, each country will be required to draw on its own unique potentiality. Accordingly, the following remarks outline a few conjectural criteria worthy of further exploration, if only to inform and give perspective to public officials in different national contexts grappling with the challenges of ecological modernisation.

The apparent ability of the Germans and the Dutch to initiate a nascent process of ecological modernisation motivates consideration of some key characteristics shared by these two countries. First, it is useful to focus on the political sphere where we find some notable commonalties. Both countries have long traditions of fostering cooperation to resolve conflict and publics that impart high levels of legitimacy on their governments. Such qualities have enhanced civil society, promoted effective governance, and expedited the achievement of workable solutions to thorny matters of public policy (Weale 1992; Hajer 1995; Bressers and Plettenburg 1997; see also Putnam 1993). Though recent circumstances raise questions about the future durability of these arrangements, this propensity has received expression in the form of numerous partnership-building practices (e.g., roundtables, discourse coalitions) designed to facilitate consensus among industry, government, and citizen groups on critical environmental issues.

Second, we can attribute early success moving toward ecological modernisation in Germany and the Netherlands to these countries' broad public support for planning, as well as welfare liberalism more broadly. The Dutch National Environmental Policy Plan (NEPP) of 1989 is a good example of how this capability has been brought to bear on the process of ecological modernisation. It is true that detractors have criticised the NEPP in recent years (see, in particular, Opschoor and van der Stratten 1993), in part because the plan's forecasted environmental improvements have proved too ambitious. Nonetheless, the NEPP embodies an important process and the undertaking has contributed to both individual and institutional learning. More definitive confirmation of the plan's effects on environmental sustainability would require a detailed evaluation of the specific dialectic between national planning and ecological modernisation. In the absence of such a case study it is plausible to surmise that these exercises have played a central role in upgrading national capability for environmental reform.

Finally, Germany and the Netherlands rank among the world's most environmentally committed countries. It is their strong and persistent public ecological consciousness that has driven demand for stringent regulatory controls and created the context in which initiatives concordant with ecological modernisation could be cultivated. Surely it remains to be seen whether these attitudes can be sustained. However, the institutionalisation of the Green Party in Germany and the urgency imparted by population density in the Netherlands suggest that environmental values in these nations may not be subject to the erratic swings commonly found elsewhere.

This discussion then leads to the judgement that countries that lack the aforementioned attributes will have difficulty pursuing ecological modernisation. While such deficits do not preclude nations from embracing this reform programme, candidate countries will need to fashion some means of compensating for these deficiencies.

References

Andrews Richard NL. 'United States.' In *National Environmental Policies: A Comparative Study of Capacity Building*, Martin Jänicke and Helmut Weidner, eds. Berlin: Springer, 1997.

Ayres Robert U. *Industrial Ecology: Towards Closing the Materials Cycle*. Cheltenham: Edward Elgar, 1996.

Andersen Mikael S. *Governance by Green Taxes: Making Pollution Prevention Pay*. Manchester: Manchester University Press, 1994.

Bressers Hans TA, Plettenburg Loret A 'The Netherlands', In *National Environmental Policies: A Comparative Study of Capacity Building*, Martin Jänicke and Helmut Weidner, eds. Berlin: Springer, 1997.

Brundtland Gro Harlem. The challenge of sustainable production and consumption patterns. *Natural Resources Forum* 1994; 18(4): 243 - 246.

Caincross Frances. *Costing the Earth: The Challenge for Governments, the Opportunities for Business*. Boston: Harvard Business School Press, 1992.

Christoff Peter. Ecological modernisation, ecological modernities. *Environmental Politics* 1996; 5(3):476-500.

Cohen Maurie J. Risk society and ecological modernisation: alternative visions for postindustrial nations. *Futures* 1997; 29(3):105 - 119.

Cohen Maurie J. 'Evidence of new environmental ethic: assessing the trend toward investor and consumer activism.' In *The Role Business Ethics in Economic Performance*, Ian Jones and Michael Pollitt, eds. Basingstoke: Macmillan, 1998a.

Cohen Maurie J. 'Science and society in historical perspective: implications for social theories of risk. ' In *Risk in the Modern Age: Contributions for the Social Sciences*, Maurie J. Cohen, ed. Basingstoke: Macmillan, 1998b.

Commoner Barry. *Making Peace with the Planet*. New York: Pantheon, 1990.

Daly Herman E. *Beyond Growth: The Economics of Sustainable Development*. Boston: Beacon Press, 1996.

Dowie Mark. *Losing Ground: American Environmentalism at the Close of the Twentieth Century*. Cambridge: MIT Press, 1995.

Durning Alan Thein. *How Much is Enough? The Consumer Society and the Future of the Earth*. New York: W. W. Norton, 1992.

The Economist. Business-Greenpeace: A Model for Business? 336 (7928):61, 1995.

Fischer Kurt and Schot Johan. *Environmental Strategies for Industry: International Perspectives on Research Needs and Policy Implications*. Washington, DC: Island Press, 1993.

Gore Al. *Earth in the Balance: Forging a New Common Purpose*. New York: Houghton Mifflin, 1992.

Gouldson Andrew and Murphy Joseph. Ecological modernisation and the european union. *Geoforum* 1996, 27(1):11 - 21.

Hajer Maarten. Ecological modernisation as cultural politics, In *Risk, Environment, and Modernity: Towards a New Ecology*, Lash Scott, Szerszynski Bronislaw, and Wynne Brian, eds. London: Sage, 1996.

Hajer Maarten. *The Politics of Environmental Discourse: Ecological Modernisation and the Policy Process*. Oxford: Clarendon Press, 1995.

Harvey David. *Justice, Nature, and the Geography of Difference*. Oxford: Blackwell, 1996.

Hays Samuel P. *Conservation and the Gospel of Efficiency: The Progressive Conservation Movement, 1890-1920*. Cambridge: Harvard University Press, 1959.

Huber Joseph. *Die Regenbogengesellschaft: Okologie and Sozialpolitk (The Rainbow Society: Ecology and Social Politics)*. Frankfurt am Main: Fischer, 1985.

Jänicke Martin. *Preventive Environmental Policy as Ecological Modernisation and Structural Policy*. Berlin: WZB, 1985.

Jänicke Martin and Weidner Helmut, eds. *National Environmental Policies: A Comparative Study of Capacity Building*. Berlin: Springer, 1997.

Jänicke Martin, Mönch Harald, Ranneberg Thomas, Simonis Udo E. Economic structure and environmental impacts: East-West comparisons. *The Environmentalist* 1989; 9(3):171-183.

Makower Joel. *The E-Factor: The Bottom-Line Approach to Environmentally Responsible Business*. New York: Random House, 1993.

Marsh George Perkins. *Man and Nature*, edited by David Lowenthal. Cambridge: Harvard University Press, 1965 [1864].

Mol Arthur P. J. *The Refinement of Production: Ecological Modernization Theory and the Chemical Industry*. The Hague: CIP Data Koninklijke Bibliotheek, 1995.

Moore Curtis and Miller Alan. *Green Gold: Japan, Germany, the United States, and the Race for Environmental Technology*. Boston: Beacon Press, 1994.

Myers Norman. Consumption: challenge to sustainable development or distraction? *Science* 1997; 276(4 April):53-57.

Opschoor Hans and van der Straaten Jan. Sustainable development: An institutional approach. *Ecological Economics* 1993; 7(3):203 - 222.

Organisation for Economic Cooperation and Development. *Environmental Performance in OECD Countries: Progress in the 1990s*. Paris: OECD, 1996a.

Organisation for Economic Cooperation and Development. *Environmental Performance Reviews: United States*. Paris: OECD, 1996b.

Organisation for Economic Cooperation and Development. *Environmental Performance Reviews: The Netherlands*. Paris: OECD, 1995.

Organisation for Economic Cooperation and Development. *Environmental Performance Reviews: Germany*. Paris: OECD, 1993.

Palmer Karen, Oates Wallace E., Portney Paul R. Tightening environmental standards: The benefit-cost or the no-cost paradigm? *Journal of Economic Perspectives* 1995; 9(4):119 - 132.

Pearson Charles S., ed. *Multinational Corporations, Environment, and the Third World*. Durham: Duke University Press, 1987.

Porter Michael E. and van der Linde Claas. Toward a new conception of the environment - competitiveness relationship. *Journal of Economic Perspectives* 1995a; 9(4):97 - 118.

Porter Michael E. and van der Linde Claas. Green and competitive: Ending the stalemate. *Harvard Business Review* 1995b; 73(5):120-134.

Porter Michael E. America's Green Strategy. *Scientific American* 1991; 268 (April):96.

Porter Michael E. *The Competitive Advantage of Nations*. London: Macmillan, 1990.

Portney Paul R. Green and Competitive: Response. *Harvard Business Review* 1995; 73(6):199 - 200.

Putnam Robert D. *Making Democracy Work: Civic Traditions in Modern Italy*. Princeton: Princeton University Press, 1993.

Redclift Michael. *Wasted: Counting the Costs of Global Consumption*. London: Earthscan, 1996.

Rostow WW. *The Stages of Economic Growth*. Cambridge: Cambridge University Press, 1960.

Sachs Wolfgang, ed. *Global Ecology: A New Arena of Political Conflict*. London: Zed Books, 1993.

Schmidheiny Stephan. *Changing Course: A Global Business Perspective on Development and the Environment*. Cambridge: MIT Press, 1992a.

Schmidheiny Stephan. The business of sustainable development. *Columbia Journal of World Business* 1992b; 27(3 - 4):18 - 24.

Schmidheiny Stephan and Zorraquín Frederico. *Financing Change: The Financial Community, Eco-Efficiency, and Sustainable Development*. Cambridge: MIT Press, 1996.

Shrivastava Paul. Democratic control of technological risks in developing countries. *Ecological Economics* 1995; 14(3):195 - 208.

Simonis Udo E. *Beyond Growth: Elements of Sustainable Development*. Berlin: Edition Sigma, 1988.

Socolow Robert H., ed. *Industrial Ecology and Global Change*. Cambridge: Cambridge University Press, 1994.

Walley Noah and Whitehead Bradley. It's not easy being green, *Harvard Business Review* 1994; 72(3):46 - 52.

Weale Albert. *The New Politics of Pollution*. Manchester: Manchester University Press, 1992.

Weizsäcker Ernst Von, Lovins Amory, Lovins Hunter. *Factor Four: Doubling Wealth, Halving Resource Use: The New Report to the Club of Rome*. London: Earthscan, 1997.

Welford Richard. *Corporate Environmental Management: Systems and Strategies*. London: Earthscan, 1996.

World Commission on Environment and Development. *Our Common Future*. Oxford: Oxford University Press, 1987

6 LOCALIZED TECHNICAL CHANGE AND THE EFFICIENT CONTROL OF GLOBAL WARMING

Francesco Ferrante**

Introduction

The purpose of this chapter is not to suggest an explicit ranking of different pollution control instruments by dynamic cost-effectiveness, nor to analyze each instrument in detail. Rather, this work first points out some methodological limitations of the techniques used to date to model environmental policy scenarios in the energy sector. Building on this, it then identifies the features that characterize cost-effective, politically feasible strategies to curb global warming in the presence of institutional inertia due to the cost of technological adjustment. I believe that this investigation helps to explain the substantial failure of both national and international efforts at co-ordinated emissions control strategies relying on fossil fuel or energy taxes.

The core of this analysis is an investigation of the determinants of the direction of technical change and of the effects of environmental regulation on the incentives for environment-saving innovation. The state of the discussion on environmental regulation, technical change and productivity growth - which has recently focused on the 'Porter hypothesis' (Porter and van der Linde 1995; Palmer, Oates and Portney 1995; Jaffe and Palmer 1996) – does not offer a neat picture of the casual linkages. The few studies available suggest that the channel connecting environmental regulation and dynamic economic performance is

** Università di Cassino, Dipartimento Economia e Territorio, Cassino (FR) 03043, tel. 0776-299424; e-mail ff@serv.eco.unicas.it. The financial support from MURST is gratefully acknowledged.

crucially affected by industry- and firm-specific costs that are technological in nature. This suggests that our theoretical models should explicitly factor such costs in. Most important, policy makers should be aware of these constraints.

The economic intuition underlying the analysis developed here is that in the absence of markets for the emissions, technical change has generated an environmentally detrimental bias in the allocation of energy R&D efforts. Postulating that technological advance is localized implies that today's decisions on the allocation of R&D efforts for emissions abatement are powerfully affected by a severely biased distribution of opportunities between conventional and clean energy technologies that ultimately favours the former. This limits the role of markets as devices for efficient allocation of R&D; in particular, I argue that in the absence of appropriate policy action, localized technical change will delay the development of cleaner energy technologies and foster the diffusion of energy consumption patterns worse than those that are socially efficient in the long-run.

Methodologically, the chapter further shows that on quite reasonable assumptions on the interdisciplinary structure of energy research, localized technical change can be considered a significant source of policy-relevant non-convexities in the set of production possibilities and that it also offers a convincing explanation of the institutional, scientific and market barriers to eco-efficient technology that are generally said to operate in the energy sector. By barriers to eco-efficiency I mean all those 'factors creating unexplored opportunities for increased cost-effectiveness in the provision of energy services in specific applications' (Sanstad, Koomey and Levine 1993, p. 2).

The main conclusion is that the first need in designing energy regulatory mechanisms for environmental control purposes is to identify the opportunities for non-price-induced change (the reference is obviously to fuel prices) in the patterns of energy production and consumption and the policy actions required to exploit them fully. In this context, the need to enlarge the set of emission control instruments to targeted R&D and technology policies is crucial.

The arguments developed here carry implications for the debate on mitigating global warming. The first methodological indication is the unreliability of the abatement-cost estimates of studies using a top-down modelling approach. Conversely, my results tell in favour of a bottom-up, forward-looking assessment of the opportunities for policy-induced technical change.

The second implication is that the agents who would bear, in the short term, most of the technological adjustment costs are induced to lobbying against the creation of markets for the emissions (through, e.g., energy taxation). Considerations of political feasibility, accordingly, imply that localized technical change would require measures to *spread* the burden of pollution control and, hence, reduce the extent of institutional inertia.

This chapter is organized as follows. Section 6.1 sets out an operational definition of localized technical change and discusses its impact on the expected technological evolution of the energy systems in the absence of markets for the emissions. Next, the analysis focuses on the role of localized technical change as a potential source of non-convexities and institutional inertia. Section 6.2 derives the principal methodological and policy implications of localized technical change with respect to global warming. Section 6.3 draws the conclusions and lists issues that deserve to be included in a practical research agenda on global warming.

6.1 Technical Change as a Localized Process

Atkinson and Stiglitz (1969) and David (1975) have sketched the main properties of processes characterized by *localized* technical advance. The framework presented here is not intended to model a *realistic* scenario faced by the firm/innovator in a dynamic decision making context but, less ambitiously, to show the kind of impact that localized learning at micro level may have on scientific and technological evolution at macro level. Yet, it seems to me that this simple model is consistent with a wide range of behavioural assumptions about the actual motivations and strategies of firms.

As Atkinson and Stiglitz (1969) suggest, economists seem to have a short memory in that they often forget that the neoclassical representation of the set of production possibilities as a smooth differentiable curve is just an approximation. The different points on the curve do in fact represent different production techniques characterized by a given scientific and technical content of knowledge. The key implication of this non-trivial qualification is that the firm's decision on the optimum input combination at a given point in time also implies a choice between production processes. Moreover, moving from static to dynamic issues, this qualification requires that in order to analyze technological change one should restrict the analysis to single points or particular neighbourhoods on the isoquants map. As a result of this, one must not assume that technological innovation will lead to a shift of the entire production function; in such a case one should postulate complete *interprocess* externalities from R&D, therefore, that the advance achieved for the technique actually used can be extended to the other techniques at negligible cost. This is a very strong assumption indeed, in that technical change is a *problem solving activity* that is driven by prior decisions to use some particular technique within the set of the options available (David, 1975; Atkinson and Stiglitz, 1969). In this respect, innovation is essentially and typically a selective, path-dependent process.

In order to characterize an economic environment displaying these characteristics, I represent the set of the firm's production possibilities in the standard way as the relationship between inputs and output. The minimum amount of inputs necessary to produce 'E' units of output is given by the set G(E), which is defined as:

$$G(E)=\left\{Z \in R_{+}^{2} : (Z,E) \in E \text{ and } (Z,E') \notin E \ \forall E' > E\right\}$$

where

R_{+}^{2} is the two-dimensional set of positive numbers;

E is the output flow

$Z=(X,F)$ is the inputs vector

Figure 6.1 depicts the available set (E0) of an ideally infinite number of techniques or production processes available to generate one unit of 'E' at time 0.

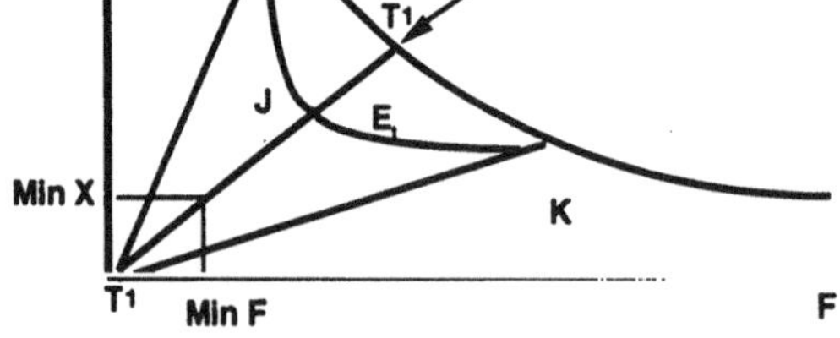

Figure 6.1 Localized technical change

Let us assume that there are 'M' such techniques and that 1/wx is the quantity of input 'X' per unit of output and 1/wf is the quantity of input 'F' per unit of output using technique 'm' at time 't'. The input requirement set per unit of output is given by the following:

$$G(1)=\left[\left(\frac{1}{w_{xt}^{1}},\frac{1}{w_{ft}^{1}}\right),\ldots,\left(\frac{1}{w_{xt}^{m}},\frac{1}{w_{ft}^{m}}\right),\ldots,\left(\frac{1}{w_{xt}^{M}},\frac{1}{w_{ft}^{M}}\right)\right] \tag{6.1}$$

where m=1,..i,..,M. Therefore, $w^{m}xt$ and $w^{m}ft$ are efficiency parameters that reflect the state of technology at time 't'. Now, the quantity of inputs 'X' and 'F' is fixed and the different techniques are generated by pairs of technical coefficients (wx, wf). If we assume constant returns to scale, the input requirement set for any level of 'E' is given by:

$$G(E)=\left(E_{\phi_1},\ldots,E_{\phi_m},\ldots,E_{\phi_M}\right)$$

$$\text{where } \phi_m=\left(\frac{1}{w_{xt}^{m}},\frac{1}{w_{ft}^{m}}\right) \tag{6.2}$$

G(E) is the isoquants map for this set of technological possibilities. I assume that firms are cost - minimizers, therefore, that they choose the production process that minimizes costs given their expectations on input prices. Hence, given $(px, pf)^{e}$ firms adopt a technique (production process) 'ϕi' that satisfies:

$$\phi_i=\left\{\left(\frac{1}{w_{xt}^{i}},\frac{1}{w_{ft}^{i}}\right):\left(\frac{p_x}{w_{xt}^{i}}+\frac{p_x}{w_{ft}^{i}}\right)<\left(\frac{p_x}{w_{xt}^{m}}+\frac{p_x}{w_{ft}^{m}}\right)\forall\, m\in M\right\} \tag{6.3}$$

where, m=1,..,i,..,M is the set of techniques. This means there will be 'M' ex ante techniques, all candidates for the role of *core* techniques depending on the prevailing expectations on input prices. A *localized* process of technological upgrading in the input space can be represented as I did in Figure[1] 6.1. The main feature of localized technical change is that the advance achieved for the *core*

[1] Following the convention, the set of techniques is represented as a continous curve.

technique over time can be used to ameliorate the other techniques only after adaptation and, in any case, at a reduced pace. Presumably, the feasibility and opportunity of such adaptations diminishes as the distance separating each technique from the *core* increases. In fact, this technological distance reflects the degree of technological homogeneity and scientific similarity between different techniques in the input space. Accordingly, the direction of technical change should have a significant impact also on the direction of scientific advance, in cumulative fashion. Thus, the latter process reinforces the *localized* nature of technical advance[2]; notably, it increases the idiosyncrasy of the investment in R&D. That is to say, it raises the cost of technological adaptation over time.

Technical change can be measured in terms of the rate of increase of w_x and w_f over time. Let us assume that $\omega^{ct}(RD^*, \theta_{ct})$ is the rate of *locally neutral* technical[3] change for the core technique at time 't', where θct is a parameter of *technological opportunity* that characterizes the core technique, therefore, it determines the productivity of R&D (with $\partial\omega/\partial RD^* > 0$, $\partial\omega/\partial\theta ct > 0$) and δ_{cm} is the distance (e.g. the Euclidean distance at the *reference point*[4]) between the technique 'm' and the core technique. I posit that RD* is the amount of R&D chosen by the innovator/firm[5]: in a world of optimizing firms, one should expect that RD* is determined by the condition that the expected marginal benefits from R&D are equal to the marginal costs[6]. Taking the price of the inputs $(px, pf)^e$ as given, the *locally neutral* rate of technical change enjoyed[7] by each *ex-ante* technique $m \in c$ is:

$$\omega^{mt} = \sigma(\omega^{ct}(RD^*, \theta ct, t); \delta cm) \qquad (6.4)$$

where: $\sigma(RD^*, \theta ct, 0,) = \omega^{ct}$; $\partial\sigma/\partial RD^* > 0$, $\partial\sigma/\partial\delta < 0$, $\partial^2\sigma/\partial\delta^2 > 0$ $\partial\sigma/\partial\theta ct > 0$, $\partial\sigma/\partial t < 0$ $\forall$ m=1,..,M. Moreover, $\sigma((\omega^{ct}(RD^*, \theta ct, t); \delta cm > \delta°) = 0$ or $\sigma((\omega^{ct}(RD^*, \theta ct, t); \delta c(j \in J)) = 0$ where $J \subseteq M$ is the set of techniques that do not enjoy

---

[2] Rosemberg (1982), Dosi (1982) and Dosi (1988) lend empirical support to this hypothesis.

[3] That is to say, $\omega^{ct}(RD^*, \theta ct,)x = \omega^{ct}(RD^*, \theta ct,)f$ the rate is the same for the two inputs.

[4] It may be reasonable to assume that this technological distance changes over time as a result of technical advance.

[5] In a standard neoclassical model, R&D* would be determined as the profit-maximizing value of R&D expenditure.

[6] Given the cost-minimising decision taken at any point in time, localized technical change can be obtained under quite general behavioral assumptions about the process of innovation decisions at firm level.

[7] I.e., the rate of change can be transferred from technique those 'c' to technique 'm'. Local neutrality implies that the one moves along a vector toward the origin.

any technological spillover. Hence, *given the choice of the technique carried out at the reference point,* the set of production possibilities in each period 't' is given by:

$$G(E)_t = \left(E_{\phi 1t},\ldots,E_{\phi mt},\ldots,E_{\phi Mt}\right)$$

$$\text{where } \phi mt = \left(\frac{1}{e^{t\sigma^m}\,w^m_{x0}},\frac{1}{e^{t\sigma^m}\,w^m_{f0}}\right) \tag{6.5}$$

This characterization of the technological environment implies that only a subset of techniques 'S' that belongs to an appropriate neighbourhood of the *core* technique is up-graded over time. Moreover, the pace of advance enjoyed by those techniques that belong to the technological spillover region is smaller than that achieved by the *core* technique and displays a time-decreasing trend. As a consequence, the *ex-post* set of production possibilities will display a time-varying increasing curvature around the *core* technique. The final impact of this *localized* and *locally neutral* pattern of innovation is represented in Figure 6.1 through the movement and change in the shape of the unitary isoquant (E0 -> Et).

In this simple setting, a reasonable account of the evolution of the technological opportunities, which are assumed to be the main source of exogenous technological advance, might be based on the idea that the latter are determined by the pattern of scientific advance. Going by this plausible explanation, then, and to the extent that there is a feedback mechanism linking[8] the generation of technological change at micro level with the production of scientific knowledge at macro level (Ferrante, 1996), one can realistically expect the *localized* pattern of technical change to be reinforced by the process of scientific advance.

6.1.1 Localized Technical Change and Missing Markets for the Environment in the Energy Sector

For simplicity, let us assume now that at time '0' – our reference point – 'E' is the unitary isoquant for the generation of energy, therefore, it represents all the available technological combinations between input 'X' and input 'F' yielding one unit of energy. 'F' is a polluting fuel, e.g. a fossil fuel, and 'X' is an emission-free input. Moreover, let us suppose that, on the basis of the price of

[8] On this point see e.g., Dosi (1988) and Sahal (1981).

the polluting fuel and the absence of markets for the polluting emissions, a cost-minimizing but emissions-intensive technique is chosen ('T' in Figure 6.2). As noted, owing to localized technical change, the choice of a cost-minimizing technique at the reference point also implies a decision on the innovation strategy. In the case of adaptive[9] expectations about the relative prices of the fuels and no markets for the emissions, the direction of innovation would yield a locally neutral pattern of technical advance along the ray TT' in Figure 6.2.

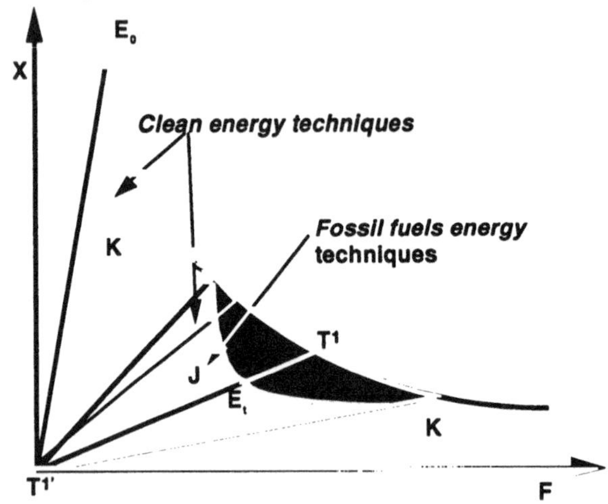

Figure 6.2 Localized technical change and missing markets for the emissions in the energy sector

This process is locally neutral with respect to the set of techniques within the neighbourhood of those actually used but globally biased with respect to the complete spectrum of techniques available *ex ante*. As we have seen, this is because innovators would develop capabilities to upgrade the core fossil fuel combustion technique and the techniques in its neighbourhood (e.g. coal, oil, gas burning energy generation techniques) that share its scientific principles.

If the expected prices of the different fuels do not change for a long time, overlapping generations of techniques will be used and advanced that are *locally neutral* but *globally biased* against low-emissions energy techniques. Given the characterization of this economic environment and the shape of the production

---

[9] The assumption of *rational expectations* would not change this result unless one supposed that innovators anticipate an institutional innovation such as the implementation of markets ofr the environment in the future.

possibilities set, the main policy-relevant considerations deriving from the effects of this localized pattern of innovation are the following:

(a) The increase in technological adjustment costs reduces the substitutability among techniques and, consequently, among energy fuels;

(b) the private as well as the social opportunity cost of shifting the direction of technical change towards clean generating and end-use energy techniques increases over time;

(c) due to the generation of environmental and technological externalities, the social opportunity costs (benefits) of changing the direction of technical change are smaller (larger) than the private cost.

Of course, this is no mere hypothetical account. On the contrary, it seems to provide a realistic description of the evolution of the energy system over the last 150 years or so and of the economic and technological constraints that characterize it today. Industrialization relied on the massive exploitation of cheap fossil fuels all over the world. Until the 1920s the *core fuel* was coal. In the last 70 years or so, it has been progressively supplanted by oil with its lower price in energy-efficiency units.

If this is the actual story, it is not difficult to imagine the impact of localized technical change – according to (a), (b) and (c) – on the private and social cost of controlling emissions today by adjusting fuel prices. Owing to sunk costs, the generation that decides to bear the cost of emissions control through Pigouvian pricing (pollution taxes or permits) would face substantial welfare losses in consumption of conventional goods. In fact, given high technological adjustment costs, pollution control through artificial markets alone would work most of its effects by means of output adjustments.

6.1.2 Localized Technical Change as a Potential Source of Non-convexities

In the form set forth above, localized technical change is expected to generate a single non-convexity problem. Figure 6.3 shows that as a result of a localized pattern of technical change, which has excluded from the improvements most of the latent techniques, given the input price ratio pe/pf , there is a multiple equilibria problem. The assumption behind this rather specific representation of the process of technical progress is that the degree of technological and scientific homogeneity between the *core* and the other techniques is monotonically related to the distance between them in the input space. But this is an analytical simplification that cannot claim general validity. The links between scientific disciplines and between research areas cannot be uniquely related to the technological characteristics of any given input combination. Most important, the opportunities to extend the advance in the *core* techniques to others over time, cannot be expected to be monotonically related to said distance. Still, one can nevertheless argue that the actual set of techniques enjoying these scientific and technological spillovers is smaller than the *ex ante* set of techniques and that moving very far away from the *core* such spillovers are nil.

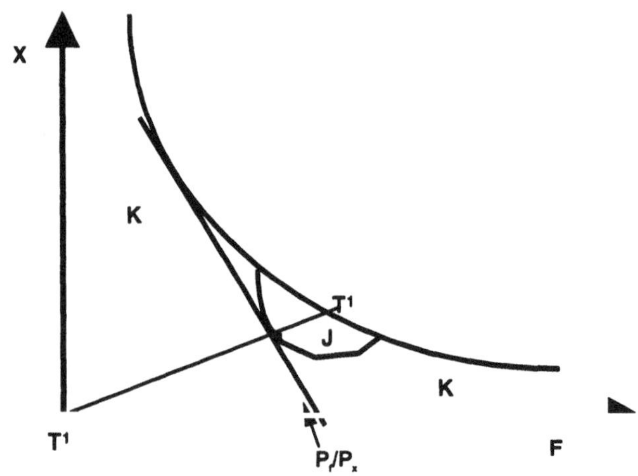

Figure 6.3 Localized technical change as a source of nonconvexities

From these assumptions it follows that the actual shape of the unitary isoquant will not be single-peaked but will present a number of local peaks around that of the *core* technique[10]. To simplify matters – but without loss of generality – let us assume that the transferability of spillovers from the *core* technique to each technique 'm' depends on the characteristics of the latter and that the potential for spillovers depends on the value of a technological opportunity parameter θmt. Then,

[10] The rationale behind this representation lies in the analysis of the interdisciplinary structure of research. An interesting example of the network of interdisciplinary links in energy research between different scientific and technological fields and of the potential for spillovers is provided by Tijssen (1992).

$$\omega^{mt} = \sigma(\omega^{ct}(RD^*, \theta ct, t); \delta cm, \theta mt) \qquad (6.6)$$

where: $\partial\sigma/\partial RD^* > 0$, $\partial\sigma/\partial\delta < 0$, $\partial\sigma/\partial\theta ct > 0$, $\partial\sigma/\partial\theta mt > 0$, $\forall\ m=1,..,M$. Moreover, $\sigma(\omega^{ct}(RD^*, \theta ct, t); \delta cm > \delta°, \theta mt) = 0$ or $\sigma(\omega^{ct}(RD^*, \theta ct, t); \delta c(j \in J); \theta mt) = 0$.

Leaving aside negative externalities such as pollution, in a neoclassical world of optimizing agents non-convexities are potentially serious sources of market failure. Prices do not work and the pursuit of economic efficiency may require policy intervention and alternative resource allocation devices.

The crucial point for the policy maker is gathering information about the likely location of the peaks in the social transformation curve: "The problem really concerns the availability of global information. If the government really knew the technological and economic possibilities in their entirety (...) there would be no problem (...) typically the net benefit schedule over the entire non-negative range of pollution is not known to the government. The regulator picks up the problem at the market equilibrium point. Typically, it contemplates marginal moves (it obtains only local information about the net benefit curve from the firms) and continues supporting these moves as long as the net benefits increase at each move" (Dasgupta and Heal, 1993, p. 91).

Henceforth, in the presence of technological non-convexities, the search for dynamic efficiency should be guided by *global* cost-benefit analyses going beyond the local spectrum of technological options.

The potential for the generation of technological non-convexities and, therefore, the need for policy intervention in basic research is heightened by the fact that the link between science and technology may be one of cumulative causation: "Undoubtedly, the interaction between fields of technology and between science and technology is often characterized by many feedback processes. Consequently, interactions will be of 'non-linear' type. This is an important conclusion because non-linearity may imply, as we know from recent developments in physics and mathematics, a chaotic, therefore, an unpredictable behaviour" (Engelsman and van Raan 1994, p. 1).

Under quite reasonable assumptions about the interdisciplinary structure of science and technology one should suspect that localized technical change may have generated significant non-convexities in the field of energy technologies. This implies that policies for the efficient control of global warming should be concerned primarily with accurate investigations, therefore, cost-benefit analysis applied to energy R&D projects, of the likely location of the *global optimum*. Such studies should seek to identify the most promising energy technologies, the cost of developing them to their full thermodynamic efficiency (over a suitable period of time) and their potential market penetration under alternative economic and policy scenarios. This task requires a good deal of co-operation between energy analysts and economists.

6.1.3 Localized Technical Change as a Potential Source of Institutional Inertia

Two main potential sources of institutional inertia are inherent in the localized nature of technical change. The first stems from the political pressure of the socio-economic groups that have invested in the *core* techniques. These groups, who would be affected by the redistribution of the property rights on the environment, have a powerful incentive to block or delay this process, which substantially favours future generations and environmentalist groups, in other words, those groups which attach a high value to conservation and environmental quality. Unfortunately, owing to organizational failures, micromotives cannot simply be aggregated to analyze macro behaviour: the main aspect of this account worth emphasizing here is that, depending on members' expected cost/benefit ratio and the size of the group (Schelling, 1978), free-riding may block the formation of some of the latent groups. The incentive of individuals to coalesce into pressure groups is determined by two contrasting forces: the expected gains from exerting political influence and the costs borne by individuals to organizing themselves as lobbies (set up costs). Most important, if one takes proper account of the micromotives of different individuals in the environmental policy arena, one should expect that, because of inadequate representation, the ***equilibrium*** level of pollution control resulting from political competition will generally be socially sub-optimal. This sub-optimal equilibrium resulting from the political market should not be ascribed, as is usually done, only to the non-representation of future generations but also to poor expected cost/benefit ratio and, hence, to a low participation rate of the individuals who attach high values to environmental conservation.

A second potential source of institutional inertia is the setting of technological standards[11], which naturally follows and drives the process of technological advance: "Standardization as an intentional act or as an unintended consequence of the interplay of actions taken for other purposes, creates order by reducing variety. Order offers reduced uncertainty and thus permits economizing on the costly gathering of information (...) order may serve to focus experimentation in

[11] Standards may be established by law (*de jure*) or by practice (*de facto*).

Figure 5.3B Indexed GDP and selected emissions to Air Japan, 1980 - 1994 (1980 = 100)

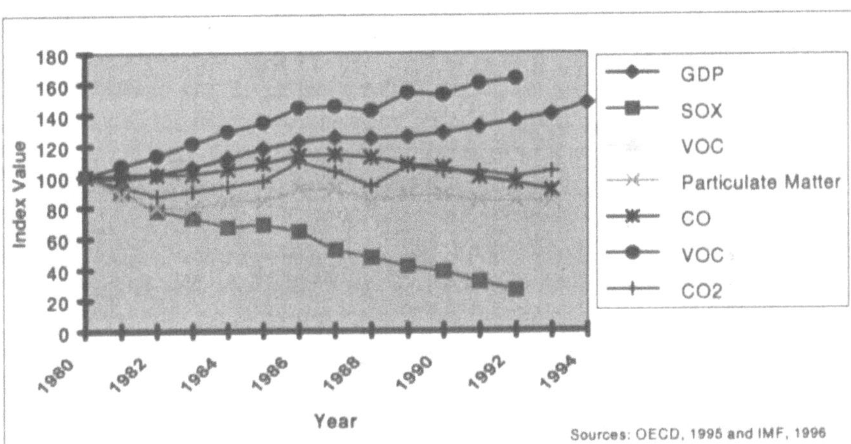

Figure 5.3C Indexed GDP and selected emissions to Air Norway, 1980 – 1994 (180 = 100)

The data on emissions to air indicate that the other candidate countries have not yet begun to disengage their economies from the environment. Such evidence implies that there has been little quantifiable progress toward ecological modernisation in these nations and their modern economic structures remain intact. Data for the United Kingdom, Norway, and Japan are presented in Figures

At first glance, the model seems to support the top-down modelling approach and the conclusion that, due to high technological transition costs, there is no *free lunch,* therefore, no unexplored opportunities for eco-efficiency. This is just a *prima facie* argument; it is easy to verify that, on the assumptions of the model there is scope for cost-effective policy intervention. This means (a) that there are market co-ordination failures that generate inefficiencies; (b) that there is a policy-relevant gap between the social and the private rate of returns to energy and eco-efficiency.

The model provides a rationale for the findings of the many empirical studies suggesting that, because of market and institutional barriers, there are opportunities in the energy sector that only government action can exploit. In particular, the ideas developed above help explain technological and institutional inertia (due to adjustment costs) found by various energy studies which has been labelled as "market barriers" or "institutional barriers" to energy efficiency. Market barriers are implied by engineering calculations "when a device based on proven technology saves energy at or below the price of energy and is not being adopted after being on the market for several years. In addition, there must be no hidden costs and the input parameters used for the calculation must capture the range of possible physical situations and the usage characteristics throughout the economy" (Sanstad, Koomey and Levine 1993, p. 2). Along the same lines, market barriers to the development of eco-efficient technologies are implied when the rate of return on the R&D capital employed in an energy project is higher, according to engineering calculations, than the rate which is obtained on energy projects of equivalent riskiness and this project is not carried out for several years. There is an intrinsic linkage between market barriers, which limit the potential size of the market for innovations, and the generation of R&D barriers.

Analyses based on market barriers also suggest that price-induced changes in the structure of energy production and consumption to achieve a sustainable pattern of CO2 emissions, are not feasible, at least on political grounds, because the price changes of the magnitude required to overcome these barriers cannot be implemented in that they are neither politically feasible nor desirable. Non-price modifications, which may eventually counteract market barriers, should be considered as well.

This is the conclusion reached by Hourcade (1993), for example, who shows that, on quite reasonable modelling assumptions and various parameterizations, a low-CO2 -emissions country like France cannot rely on taxation to stabilize emissions at the 1987 level by 2000 but should implement policy-induced structural modifications in the pattern of energy consumption and production.

The same indication emerges from other analyses showing that the carbon taxes that would have to be implemented to stabilize the global emissions would be invariably very high, reaching intolerable levels early in the next century.

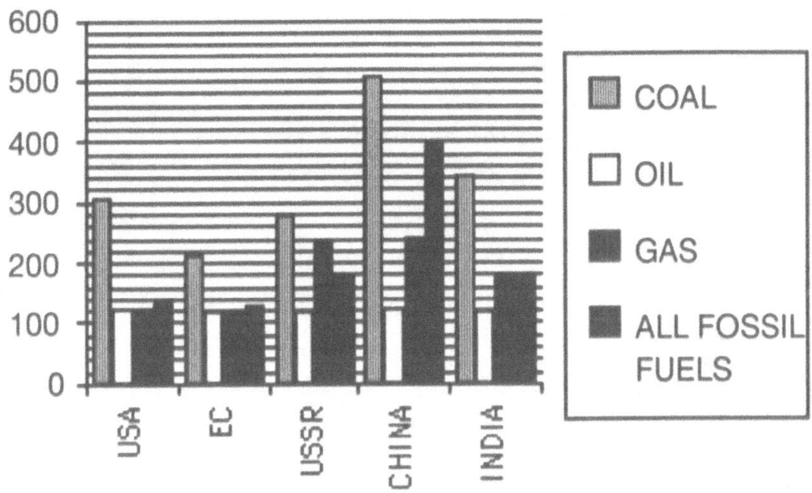

Figure 6.4 The impact of carbon taxes on end-user price by the year 2010 (price without tax =100)

Source: Haugland, Oysten and Roland, 1993

Figures 6.4 and 6.5 show the impact on the end-use price of energy 2010 of the carbon taxes required to stabilize emissions at 1987 levels (Haugland, Oysten and Roland, 1993). For example, the price of fossil fuels to end-users in China would rise by 300% (and that of coal by about 400%); the United States would face a price increase of coal of about 200% and the European Union one of about 120%. Along the same lines, the estimates presented by the European Commission estimates that a tax of $ 10 boe on all energy sources (excluding renewables) would be needed in Europe to achieve just half the abatement target required to stabilize emissions of CO2 at the 1990 level.

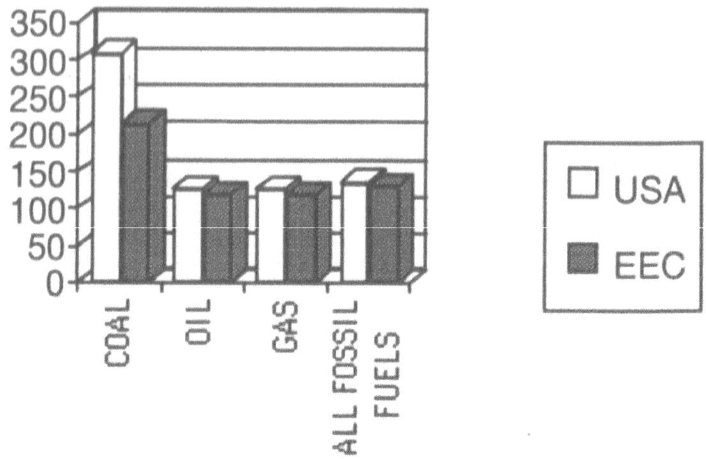

Figure 6.5 The impact of carbon taxes on end-user price by the year 2010 (price without tax =100)

Source: Haugland, Oysten and Roland, 1993

In conclusion, empirical investigation lends support to the view that due to technological adjustment costs the level of energy taxation required to achieve the emissions control targets is always very high, thus suggesting that other measures may be needed.

6.2.1 The Role of R&D and Technology Policy in the Energy Sector

R&D and technology policy are the main tools for increasing both technological and institutional flexibility. Leaving aside institutional barriers stemming from political lobbying, I have identified two major sources of market failure, viz., 'market barriers' and 'R&D barriers', which should be the target of policy intervention. The idea that there are unexplored opportunities to improve eco-efficient technologies in the energy sector is supported by several technical studies showing that the rates of return on these R&D projects are higher than on projects of equal risk (on this point, e.g. Rubin *et al.*, 1993). The emerging evidence *that there are free lunches* does not imply the strong and unacceptable conclusion that agents are irrational but simply that they lack reliable information about their opportunities and that there may be market co-ordination failures (Porter and van der Linde 1995; Sanstad, Koomey and Levine 1993). The findings strengthen the case on behalf of the idea that energy taxation alone would not suffice to make individuals alter their behaviour and that the government should adopt other measures, such as R&D and investment subsidies and the public provision of relevant basic and applied research. The list of measures is well known. Nevertheless, it is worth emphasizing the specific role of the available instruments in the pursuit of institutional and technological flexibility.

Targeted innovation and investment subsidies are required to reduce the adverse impact of market barriers to energy and eco-efficiency and, thereby, to speed up the development and adoption of clean energy technologies. By reducing the private cost of investment, they increase the incentive to develop and adopt such technologies and, hence, their potential rate of market penetration. The public provision of basic and applied research[13] aims to increase the *ex ante* knowledge in energy research, in directions that favour clean technologies. Basically, this policy approach has the same type of impact as innovation subsidies but at an earlier stage in the innovation development process. In fact, the amount of basic research available to innovators is a public good – and an input to innovation activity – that affects the productivity of R&D activity and, given the cost of the latter, the incentive to carry it out. Technology policy also includes the following tools:

(i) Performance and technology-based environmental standards

(ii) 'Environment saving oriented' procurement strategies.

(iii) The dissemination of information.

The scope of performance and technology-based environmental standards is to compel firms to adopt and develop pollution control technologies. Mandatory performance standards identify a pollution control target in terms, of say, the maximum time rate of pollutant that can be emitted, without specifying the means to achieve it. Conversely, technology-based standards identify the equipment that polluters must adopt or a significant subset of technical specifications that they must follow. Performance standards cannot be distinguished from a technology-based standard where the technical specification concerns only environmental performance (e.g. rate of emission) of a given technology. Anticipatory technology-based environmental standards deserve special mention in that they may play an important role in determining the path

[13] Including the funding of pilot projects.

of technological change (David 1995). Anticipatory standards identify a set of technical characteristics that the new technologies should possess. In the presence of localized technical change and non-convexities, they may operate as co-ordination devices to move the innovative system in the right, socially efficient direction. Moreover, in the presence of lock-in effects (Arthur, 1989), anticipatory technology-based environmental standards perform the fundamental task of forcing the development of *gateway* technologies (converters, adapters), which permit *ex post* compatibility between clean and conventional technologies, especially in the case of network technologies (e.g. transport or energy systems) or integrated technological systems.

Public procurement strategies that are 'mission oriented' to control pollution by contractual environmental standards can generate additional demand for clean technologies, strengthening the demand-pull mechanisms that trigger the innovation in environment-saving R&D. Finally, the dissemination of information, generated and collected by public agencies, performs two relevant tasks: it reduces the uncertainty surrounding the performance of untried technologies and it makes information about new technical solutions available to potential users at negligible costs.

To come to political feasibility, R&D and technology policy may contribute significantly to institutional flexibility: in fact, by redistributing the costs of changing the direction of technical change to the very large group of taxpayers, they reduce the incentive for lobbying of those groups that would otherwise have to bear most of the technological adjustment costs.

The scope for action in R&D and technology policy is even wider if one considers the international arena, where the criteria of political feasibility are more stringent. International co-ordination in implementing support measures such as. R&D subsidies appear much more realistic than the collection of a global energy tax!

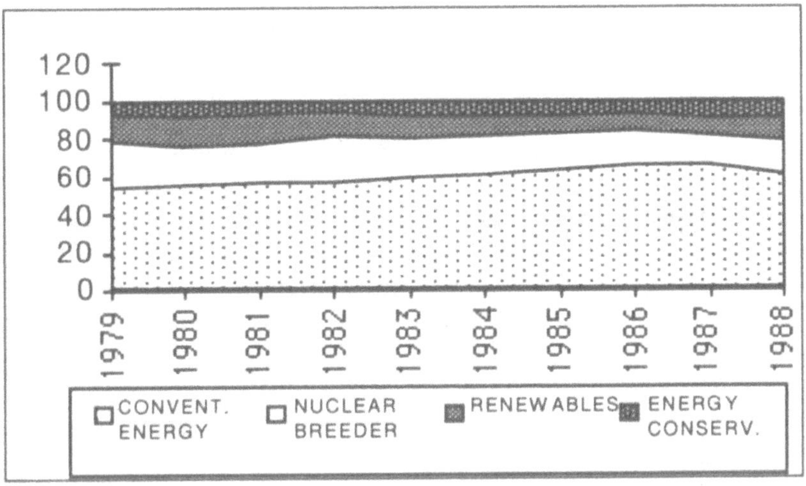

Figure 6.6 Governement funded Energy RD&D by research field (1979-1988)

Source: IEA/0ECD(1990)

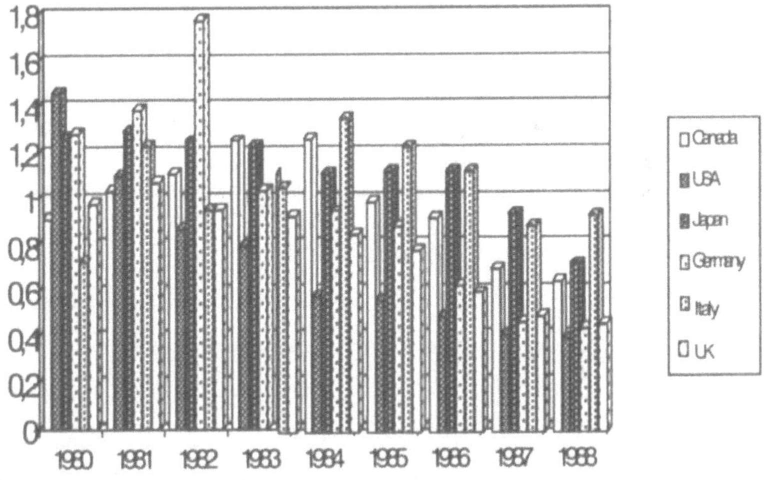

Figure 6.7 Energy RD&D expenditure per unit of GDP (%)

Source: IEA/0ECD(1990)

The data on energy RD&D (Research Demonstration & Development) spending in OECD countries indicate[14] that its potential role has been neglected so far and that the distribution of funds has favoured conventional energy technologies (the *core* techniques). Figures 6.6 and 6.7 give energy RD&D financing by the public sector in the OECD countries and its distribution by technology. Conventional technology, including conventional nuclear, on average, has received a very large slice of the cake (about 60% of the total budget). Conservation has been allocated about 12% and renewables about 8% of the budget. During years of widespread concern over global warming – with the publication of technical studies showing the existence of unexplored opportunities for energy conservation – the average expenditure destined for energy RD&D actually declined as a proportion of GDP. Most importantly, the distribution of the budget remained substantially unchanged, showing an increase, up to 1987, of the share going to conventional energy technologies. Quite likely, this pattern reflects a similar allocation pattern of private funds in energy R&D.

Of course, there is a reasonable economic explanation for this pattern, which also accounts for institutional inertia: namely, that during those years the OECD governments had their hands tied by previous commitments to support conventional energy technology programmes.

If the pattern of public expenditure in RD&D is a reflection of an underlying biased distribution of the technological opportunities in the energy sector, the data shown in Figure 6 are a good representation of the direction of technical change induced by policy intervention in the energy sector in the last decade and a half. The picture that emerges is of inter-temporal inconsistency in control strategies during the '80s: while governments were advocating transition towards eco- and energy-efficient technologies, they took actions that eventually raised the opportunity cost of taking this step in the future.

6.2.2 Top-down vs. Bottom-up Analyses of the Cost of Mitigating Global Warming

It is well known that engineers and economists disagree about the methods that should be adopted to assess the cost of abatement of emissions. Economists rely on top-down models of energy demand and supply that are calibrated over the past, whereas engineers use a bottom-up approach based on the description of

[14] Of course, these data are just a rough proxy of public involvment in energy RD&D.

the stock of technologies, of the technologies that can be economically developed and their potential diffusion to provide, over the planning horizon chosen, the same service using less energy. Hence, economists appear very much concerned with identifying market constraints to the efficient control of pollution whereas the bottom-up analysts are more interested in focusing on the technological constraints (Wilson and Wisher 1993).

One chief methodological drawback of the top-down approach is that it assumes that markets allocate investment and R&D expenditure efficiently. Under this assumption, top-down studies lead to low estimates of the net benefits of the policy actions aimed at mitigating global warming through markets: "they tell us that if it would have been expensive to reduce CO2 emissions in the past, and if the economy stays the same as it was at that time, it will also be expensive in the future" (Wilson and Swisher, 1993, p. 254). According to the present analysis, such estimates can be accepted only assuming that the induced, dynamic bias in the allocation of energy R&D, owing to lack of markets for the emissions, cannot and should not be affected by the policy action. As is argued above, however, in so doing one underestimates the opportunities for technological and behavioural adjustments driven by non-price[15] measures such as technology policy.

On the other hand, the bottom-up studies based on evaluation of the current availability of technologies underestimate the opportunities to foster the development and diffusion of new technologies that may yield negative abatement costs. In fact, as suggested, the institutional and market barriers that impede the development and adoption of energy saving technologies also reduce the private effort to develop and diffuse clean energy technologies. With regard to market barriers, various international studies suggest that there is room for significant improvement in energy conservation at *negative costs*. Economists tend to counter this optimistic view of unexplored opportunities for conservation arguing that the proponents substantially overlook the problems of the cost and the political feasibility of non-price measures. But this is not entirely so, as many studies factor in the cost of implementation of the conservation measures but still finds net benefits of policy action. For instance, taking implementation costs into account, Grubb *et al.* (1993) suggest that about 20% of energy saving and abatement activity can be conducted at negative cost to the society. The same results are obtained by Rubin *et al.* (1993), who estimated that, with discount rate values of 3%, 6% and 10%, savings of 734 BkWh (nearly 45%) could be

[15] The reference is obviously to pollution taxes and pollution permits. Technology policy may also rely on *pricing* measures such as R&D subsidies.

achieved in building electricity use in United States at a cost per kWh lower than the electricity price. This implies that significant CO2 abatement can be achieved through conservation at negative costs.

With regard to market barriers to R&D, a substantial body of studies assessing the technological potential of non-conventional energy sources lends support to the view that these sources may offer unexplored opportunities for emissions reduction with higher than markets returns on investment (see, Rubin *et al.* 1993, p .25; Mills, Wilson and Johansson, 1991, p. 540).

On practical grounds, such empirical evidence suggests that when estimating Supply Curves of Avoided Emissions (SCAE) these opportunities should be included. The resulting enlargement of the set of the technical/economic/policy options would mean a downward shift of the supply curve of avoided emissions and an increase in the expected amount of abatement that could be achieved at negative cost or positive net social benefit. Of course, the cost of policy design, implementation and enforcement should be included in order to assess the real potential for policy-driven improvements in the generation and use of energy and the actual net benefit.

Figure 6.8 shows four abatement cost curves, reflecting four different policy scenarios. The top-down cost curve (TD) lies entirely above the X-axis (therefore, has positive values); its derivation is based on the assumption that the level of emissions can be reduced only at positive cost. The standard bottom-up cost curve (BU) lies below the X-axis for energy conservation measures that can be implemented at a cost per KWh lower than the cost of electricity and above it for the other measures, including fuel switching. The extended version of the bottom-up cost curve (BURD) which includes those RD&D projects yielding a rate of return larger than the social discount rate lies also in these two orthants. Finally, the inclusion of administrative costs shifts the BURD to the new position (BURDA). Here, given the marginal abatement benefits curve (MAB) and moving from a standard top-down scenario to an extended bottom-up one, the *Pigouvian* tax required to achieve an efficient level of emissions control decreases from τ_1 to τ_2 and the efficient level of abatement increases from AE1 to AE2.

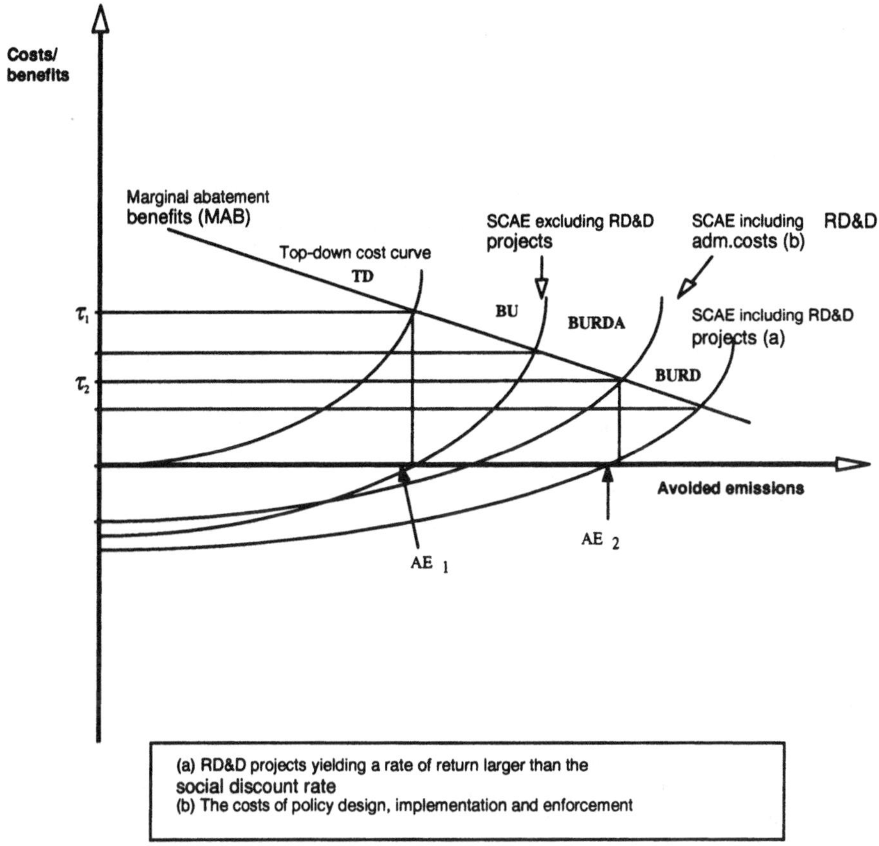

Figure 6.8 The Supply Curve of Avoided Emissions (SCAE)

6.3 Conclusions and Research Agenda

Premising realistic assumptions concerning firms' behaviour, I have shown that technical change is a localized and path-dependent process that generates technological as well as institutional inertia. On this basis, the pessimistic conclusion drawn here is that technological and institutional inertia represent serious impediments to curbing global warming by means of artificial markets for emissions. On the other hand, the analysis also offers an optimistic, pragmatic view of opportunities for mitigating global warming at low or even negative cost to society. It is argued that the exploration of these opportunities requires a mix energy taxation, R&D and technology policy – not just taxation alone. Failing such interventions a market-driven allocation of energy-related investments would tend to replicate the existing bias of technical change, which is detrimental to the environment. Furthermore, I suggest that in the presence of serious technological non-convexities generated by localized technical change, such policy action should be supported by *global* cost-benefit studies in the field of energy technologies.

Finally, some comments are warranted on the potential links between this analysis and the general concept of sustainable economic growth. The discussion has spotlighted some methodological issues. In particular, institutional as well as economic constraints appear to play a role in the analysis of long-term strategies of pollution control. The preliminary recognition of this powerful connection is a necessary methodological step, but one that environmental and energy economists have not consistently taken, at least to date. In this respect, our analysis suggests at least four main interrelated areas of research that are well worth the time and effort of investigating:

(i) The development of new models of pollution control which take explicit account of technology/institution interactions. This should include the feasibility assessment of the environmental policy instruments to be based on the analysis of the political market mechanisms.

(ii) The analysis of the interdisciplinary structures of science and of the potential for technological spillovers. This should aim to identify the *core* scientific sectors for each technological field and the potential for technological *gateways*.

(iii) The development of new methodologies for assessing the probability of technical and economic success of R&D projects in the energy sector.

(iv) Last but not least, the economics of science seems to emerge as a fundamental research area. Different forms of organization of research activity and reward systems can be expected to generate different structures of incentives and patterns of scientific advance. From the standpoint of sustainability, the pattern of advance in science and technology is a major factor indeed; one that can be ignored only at our peril.

References

Antonelli C. *Economia dell'Innovazione*. Bari: Laterza, 1995.

Arthur B. Competing technologies, increasing returns and lock-in by historical events. Economic Journal 1989; 99:116-131.

Atkinson AB and Stiglitz JE. A new view of technological change. Economic Journal 1969; 79:573-78.

Barbier EB, Burgess J, Pearce DW. 'Technological substitution options for controlling greenhouse gas emissions'. In *Global Warming: Economic Policy Responses*, Dornbusch R. and Poterba JM eds., Cambridge: MIT Press, 1992.

Binswanger HP and Ruttan V eds. Induced Innovation. Technology, Institutions And Development, Baltimore: John Hopkins University Press, 1978.

Bohm P And Russel C.S. 'Comparative analysis of alternative policy instruments'. In *Handbook of Natural Resource and Energy Economics*, Kneese AV and Sweeny J eds., Elsevier, 1985

Cline WR. *The Economics of Global Warming*. Washington: Institute of International Economics, 1992.

Dasgupta P and Heal G. *Economic Theory and Exhaustible Resources*. London: Cambridge University Press, 1993.

David P. *Technical Choice Innovation And Economic Growth*. London: Cambridge University Press, 1975.

David P. Clio and the economics of QWERTY. American Economic Review, 1985; 75(2):332-37.

David P. 'Standardization policies for network technologies: the flux between freedom and order revised'. In *Standards, Innovation and Competitiveness*, Hawkins R, Mansell R, Skea J eds., Aldershot, UK and Brookfield, US: Edward Elgar, 1995.

Engelsman EC and van Raan AFJ. A patent-based cartography of technology. Research Policy, 1994; 23:1-26.

Ferrante F. La Scatola Nera degli Strumenti di Politica Ambientale. Note Economiche, 1995; 1:103-128.

Ferrante F. Induced Technical Change, Scientific Advance and The Efficient Control of Pollution. Discussion Papers in Economics, University of York, 96/4, 1993

Grubb M, Edmonds J, Den Brinds P, Morrison M. The costs of limiting fossil-fuel CO_2 emissions: a survey and analysis. Annual Review of Energy and the Environment, 1993; 18.

Haugland T, Olsen O, And Roland K. Stabilizing CO_2. Are carbon taxes a viable option?. Energy Policy, 1993; May: 405-419.

Hope C, Anderson J, Wenman P. Policy analysis of the greenhouse effect. Energy Policy, 1993; March:327-339.

Hourcade JC. Modelling long-run scenarios. Methodology lessons from a prospective study on a low CO2 intensive country. Energy Policy, 1993; March:309-325.

I.E.A/OECD. Energy Policies of IEA Countries. 1990 Review. Paris, 1990.

Jaffe AB, Palmer K. Environmental Regulation and Innovation: a Panel Study. NBER Working Paper , N. 5545, 1996.

Jaffe AB, and Stavins RN. Dynamic incentives of environmental regulations: the effects of alternative policy instruments on technology diffusion. Journal of Environmental Economics and Management, 1995; 29:S43-S63.

Lawrence Berkeley Laboratory. The Potential for Electricity Improvements in the U.S. Residential Sector. Technical Report, LBL-30477/UC350, 1991.

Magat WA. Pollution control and technological advance: a dynamic model of the firm. Journal of Environmental Economics And Management, 1978; 5:1-25.

Mills E, Wilson D, Johansson TB. Getting started. No-regrets strategies for reducing greenhouse gas emissions. Energy Policy, 1991; July/August:526-542.

Nordhaus WD. Economic approaches to greenhouse warming. In *Global Warming: Economic Policy Responses*, Dornbusch R, Poterba JM eds., Cambridge: MIT Press, 1992.

North DC. Institutions, Institutional Change And Economic Performance. London: Cambridge University Press, 1991.

Porter ME, Van Der Linden C. Toward a new conception of the environment-competitiveness relationship. Journal of Economic Perspectives, 1995; 9:97-118.

Palmer K, Oates WE, Portney PR. Tightening environmental standards: the benefit-cost or no-cost paradigm? Journal of Economic Perspectives, 1995; 9:119-132.

Rosenberg N. Inside The Black Box: Technology and Economics. London: Cambridge University Press. 1982.

Rubin E, Cooper R, Frosch R, Lee T, Marland G, Rosenfeld A, Stine D. Realistic Mitigation Options For Global Warming. Working Paper, 1993.

Sanstad AH, Koomey JG, Levine MD. On the Economic Analysis of Problems of Energy Efficiency: Market Barriers, Market Failures and Policy Implications. Technical Report, Lawrence Berkeley Laboratory, Technical Report, LBL-32652/UC-350, 1993.

Schelling T. Micromotives and Macrobehaviour. New York: Norton, 1978.

Scheraga JD, Leary N. Improving the efficiency of policies to reduce CO2 emissions. Energy Policy, 1992; May:395-404.

Tijssen RJW. A quantitative assessment of interdisciplinary structures in science and technology: co-classification analysis of energy research. Research Policy, 1992; 21:27-44.

Wilson D, Wisher J. Exploring the gap. Top-down vs. bottom-up analyses of the cost of mitigating global warming. Energy Policy, 1993; March:249-263.

7 DO LOCAL AUTHORITIES HAVE A PART TO PLAY IN A SUSTAINABLE DEVELOPMENT?

Signe Krarup

AKF, Institute of Local Government Studies – Denmark

7.1 Introduction

During recent years the role of local authorities in relation to the environmental policy has been in focus. At the Rio conference in 1992 the importance of local involvement was also mentioned in order to assure a sustainable development. In the Danish environmental legislation this is also emphasized as local authorities are intended to play an important role in the environmental policy and are given the main authority (Miljøministeriet 1991).

The question is, however, if thinking globally and acting locally could be implemented in practice and whether this results in a better protection of the environment compared with a centralized implementation of environmental goals.

The paper focuses on the economic side of decentralization, where the environmental problems are incorporated in the general process of establishing priorities within the local authorities. First, the background for having a decentralized environmental policy will be examined, as well as which kinds of environmental protection can be appropriately decentralized. This will mainly be done from a theoretical viewpoint. Secondly, I will discuss what a decentralization of responsibilities implies for local authorities, e.g. what kind of decisions can the local authorities make in practice? I will then relate the discussion to an empirical survey about the actual role of municipalities and counties in the Danish environmental policy.[1]

7.2 A Decentralized Environmental Policy?

Even though there is a political consensus about the important role of local authorities in environmental protection there is little attention to the reasons for having a decentralized environmental protection. One can give several explanations for why a local environmental administration in some areas is better to solve environmental problems and protect environmental resources than the central administration. Or why local considerations are important in general.

Furthermore, local involvement is important because even global problems are caused by local activities. So if environmental problems in general are to be solved, this demands a change in local behaviour. The questions are therefore:

- why is it advantageous to have a local administration?
- what kind of environmental problems do local authorities have incentives to handle?

7.2.1 Differences in Needed Action

One reason for local environmental administration or local considerations is due to differences in the state of the environment across regions. This means that uniform policy goals and instruments have different impacts on the environment in different regions (Siebert 1992). In this way achieving the most appropriate protection of the environment or getting more environment for the same amount of money is difficult, as it leaves few possibilities for a selective effort.

Differences exist in the state of the environment across regions, which indicates a need for a differentiated effort. Some reasons for differences in the state of the environment could be:

a. Regional Differences in the Damage Function. Pollution, e.g. production and consumption causing emissions to the air, water and soil, has different effects on the state of the environment. The damage caused by pollution could therefore be different across regions. This indicates that the marginal damage in

1 The survey was carried out in the project Sustainability and the Possibility of Political Decisions under the research centre Society, Economics & Environment.

one region could be more severe because of a more vulnerable ecological system or a greater population density than in another region.

b. Differences in the Population's Valuation of the Environment. Most people have preferences for the state of the environment. However, people's preferences differ because of different income, occupation and lifestyle, which probably indicate a different valuation of the environment, e.g. the smell from a fishing port. This means that one region's total valuation of the environment might be different from that of another region. Big differences in the population density will also affect total valuation from region to region. In this way people and regions demand different degrees of environmental protection, and therefore a differentiated effort (Tresh 1981).

c. Regional Differences in the Costs of Abatement. The costs related to the pollution reduction might also vary across regions. This could be the case when there are regional differences in the technology of abatement or the industrial structure and when technical advantages cannot be transferred interregionally. The ability to solve an environmental problem could therefore also differ.

When a certain target for the state of a natural resource, e.g. lake, forest, soil, has to be achieved in the most effective way, the action needed at the local level will also vary across regions. To get a cost-effective solution the environmental policy and allocation of resources, one should therefore consider these differences, regardless of the level of authority. The advantages of a local administration, however, are better local knowledge of the local natural resources, local preferences, and closer contact with the polluting activity.

7.2.2 Local Problems

The environment is often characterized as a public good where the State is responsible for some degree of environmental protection. However, some functions of the environment can be characterized as a local public good. This is due to the fact that the benefits and costs of some functions of the environment have a limited geographical extension. This means that the costs and benefits associated with the solution of the relevant environmental problem or with having some type of natural resources therefore fall within the same area, such as a county or a municipality. The utility of the natural resources or the nuisance associated with pollution can in this way be limited geographically or diminish with increasing distance. This will give the local authorities an incentive to solve such pollution problems or protect such natural resources. The question whether a decentralized environmental policy is advantageous subsequently depends on the type of environmental problem under consideration.

An example: production and consumption cause emission to air, water or soil. The emission is sometimes absorbed, otherwise accumulated through pollution, in the environmental media (Pearce and Turner 1990). The limit to the environment's assimilative capacity indicates that this service function cannot be characterized as a pure, public good. When the emissions are to local or regional environmental media, it is mainly a limited geographical area that suffers from the pollution. Therefore, it is mainly the local population that benefits from reduction of such pollution problems. Pollution of a local lake, space, or noise are examples of pollution containing local costs. Clean lakes, green areas and no noise can therefore be seen as functions of the environment that can be characterized as local public goods. Here there will be local benefits connected to such goods, which give local authorities incentives to handle such environmental problems or resources. This indicates, in figure 7.1, that local incentives exist when the effects are concentrated in the local area.

		The source of pollution	
		Concentrated (point)	Dispersed (non-point)
Distribution of effects	Concentrated	Local problem	National problem
	Dispersed	National problem	National problem

Figure 7.1 Distribution of the sources and effects of pollution

Source: Naustdalslid & Hovik (1994).

However, if local action is to influence the local pollution problem the source of pollution has to be concentrated as well. If local action is to affect local problems, the source of pollution has to be local activities that affect the local environment. As local authorities are closer to the source of pollution, they might have better local knowledge than the central authorities. If local authorities have to handle pollution problems caused by local firms, they can use this knowledge. Furthermore, the residents and politicians in a municipality have a stronger incentive to engage in problems and politics concerning their own environment than the more global environmental problems. Local (environmental) goods are probably often valued higher than the national or global goods, because the local population often has a higher willingness to pay for environmental goods or the solution of local problems which more directly affect their standard of living.

The more regional or national pollution problems can also be handled by local authorities as local authorities have an important part to play if the behaviour of the local population has to be changed in a more environmentally appropriate way. However, as the local authorities do not have strong incentives for handling more national problems, an overall coordination or regulation might be needed.

To summarize, a pollution problem can be characterized, first by which kind of environmental media that is polluted, e.g. who bears the pollution costs, and second, in terms of what activities have caused the pollution problem (the source of pollution), cf. figure 1. This indicates that the pollution problem in this way determines which administrative level has the right incentives to solve the pollution problem most appropriately. Decentralization of the environmental administration therefore means that local authorities should handle local public goods as they have the incentives to do so properly.

7.2.3 The Distribution of Responsibility

If we assume that the local decision maker wants to maximize the region's total welfare, the environmental protection of local goods will be part of his general process of establishing priorities across sectors. However, if local authorities are going to solve local problems they need the ability to do so. To promote local action local authorities must be responsible for handling the more local problems.

According to Danish environmental legislation there is a clear division of authority between the Minister for the Environment, the county council and the municipal council in relation to the environmental problems concerned (Basse 1993). The overall idea in the legislation is to give local authorities the competence in environmental administration, which leaves the municipalities with the main competence. However, in two respects the competence is placed with the counties. That is when environmental problems demand more specialized knowledge or when environmental problems are interregionally transferred. However, the minister – representing the government – can always take over the competence from the counties and municipalities. For example when there are national or international considerations, when there are complicated causal relationships in the environmental sector, or when spillovers across regions or countries exist. This means for instance that:

- The municipalities have the authority in all cases of waste treatment and waste disposal. Moreover, they are in charge of the practical implementation of a cleaner technology in cooperation with the firms;

- In general, the counties manage the supervision of the ground water and surface water. However, concerning local lakes, rivers or waste water the municipalities do have the authority to supervise;
- Both counties and municipalities own part of the countryside which they have to take care of. The supervision of this is handled by the counties unless the areas considered are found in the cities or in areas with summerhouses where the municipalities have the authority;
- Each level of the environmental administration has to develop plans for the use of location space in the different geographical areas. The details in those plans increase at lower administrative levels, making municipality plans more detailed than the plan of the county as a whole. The plans also have a hierarchic structure as a municipality plan cannot be in contradiction to the county plan for the location space and
- Counties and municipalities supervise the emissions from the firms to water, soil and air, e.g. noise, waste, air emissions, waste water and obnoxious smells.

The above shows that in the Danish environmental legislation there is a connection between the environmental problems and the related administrative level as local authorities to some extent handle local or regional problems. The water resources, the soil, green areas and the air are all examples of local public goods.

However, the question is now if this means that municipalities and counties in Denmark in practice have the ability to conduct a more differentiated environmental policy. This point will be followed up in Section 7.3.

7.2.4 Problems with a Decentralized Environmental Policy

Though there are advantages to a local environmental administration, it can cause some problems when the responsibility is decentralized, which have to be considered. Local political interests sometimes take precedence over more general considerations, i.e. when activities in one community affect the welfare of other communities. The problem could therefore be that local authorities rarely have incentives to solve such problems. For example when the cost of one pollution problem in a community is small, but the total pollution has considerable effects on the environment and welfare. Or when the polluting municipality does not bear the costs of the pollution, because the pollution is transferred to other areas. If the environmental administration is decentralized, taking care of these effects in one way or the other is important. The solution can be found either through some kind of State intervention or in a way to make local authorities responsible for handling interregional spillovers. One way to take care of interregional transferred pollution could also be found by using economic instruments in the overall regulation, for example by taxation of emissions.

Local decisions can also lead to destructive inter-jurisdictional competition (Oates and Schwab 1996), where the local administration tries to attract firms by relaxing the environmental standards. This happens when the local administration favours local employment and firms by relaxing the environmental demands. However, it is uncertain how serious this problem is in practice. Commuting means that local employment often only has a limited dependence on local firms. It seems that big municipalities are less sensitive to local firms than small municipalities (Krarup 1996).

It could also be a problem that local authorities have different tasks. On the one hand, the local authorities are seen as the State's 'instrument' to implement overall policy targets, and on the other hand, they have their own local policy and targets related to local preferences and interest. Therefore, the interaction between the local and central political levels and their targets and interest is not clear, which might create regulation problems.

All of which means that, such problems with a decentralized environmental administration have to be handled if the benefits from decentralization should be fully obtained.

7.3 Which Kind of Decisions should be Handled by Local Authorities?

The local authorities in Denmark[2] are intended to play an important role in the environmental policy. What this means in practice for the local environmental protection is unclear. In order to evaluate the role of the counties and the municipalities in the Danish administration of the environment, an interview survey was carried out (cf. Krarup 1996). The empirical study was carried out through interviews with counties and municipalities of varying size and in different parts of Denmark. Through this survey it was possible to decide whether the decentralized authorities in practice have a sufficient degree of freedom to conduct a more differentiated local environmental policy.

The general findings were that there is a connection between, on the one hand, the local authorities' view and practical initiative, and on the other hand,

---

[2]Denmark consists of 14 counties and 275 municipalities, and the average population of a municipality is about 19000 people.

their size or economic situation. This means that big municipalities and municipalities without a strong budget restriction pursue a more active environmental policy. This indicates that by giving such municipalities more competence they are better able to use their knowledge about the local environment. As seen in Section 7.2.1 local authorities have the competence first of all to solve local problems and protect local resources. In this way the municipalities are able to give priority to the most severe environmental problems or most sensitive natural resources in their particular area. If this is the result, the consequence will probably be a more efficient use of resources and an improved environmental quality.

In general, environmental policy and management consist of different kinds of decisions. First, environmental policy goals have to be settled, e.g. for the state of the environment for different kinds of natural resources. Second, the implementation of such goals needs the necessary policy instruments and legislation. Third, it is necessary to supervise and control the implementation process. It is therefore interesting to discuss which administrative level should handle these different decisions most appropriately, and whether a decentralized environmental administration means that local authorities in practice take all three kinds of decisions and have the ability to make selective efforts, when the legislation gives them the competence to do so.

In the following I will discuss this in relation to the empirical study of local authorities in Denmark and give some examples. So the question is, how can local authorities solve the environmental problems they have the competence to handle, and do they have the possibility of making a differentiated effort?

7.3.1 The Setting of Targets

Setting targets for national or local environmental policy often refers to targets for the state of the environment. The State often sets national targets for reduction of emissions, waste treatment, the state of different natural resources etc. Some of these targets are transferred to the local environmental protection, e.g. for recycling and for emissions of different substances to the local environment but others, like the CO_2-emission, are rather a target for the national policy. The question is therefore whether local authorities see these targets as restrictions on their local policy or just minimum targets, trying to improve their environmental protection in some areas and setting local targets for local problems, and whether there is any room for local judgement in local environmental policy.

The possibilities for setting differentiated targets for the environmental quality across municipalities or counties depend on the environmental problem concerned. According to the survey, the municipalities do find it difficult to set their own standards for the local environmental quality affected by the firms' emissions. This is due to the so-called 'directions' made by the Minister for the Environment. These directions specify maximum allowable emissions to water, soil and air, including noise, and are specified from more general quality targets for the environmental media, where local environmental considerations are not directly included. In theory, the purpose of these guidelines is seen as a help for the local authorities in their supervision of firms. However, at the same time an appeal system exists where firms and consumers can appeal the decisions of the municipalities and counties in relation to their supervision. The practical experience with the appeal system here is that these emission standards become decisive, which indicates the difficulties the authorities have in setting more local targets. In some way, the environmental quality is therefore defined by the central authority with limited local considerations. This indicates limited possibilities for setting local and possibly differentiated environmental targets.

In areas where the municipalities have complete competence they are more able to set local targets for their environmental policy and, in some cases, set more ambitious targets than the national ones. For example, many municipalities try to improve their waste management through ambitious targets for recycling and waste treatment, and this does not come into conflict with the national targets.

As all local authorities have to develop plans for the use of location space in their area, municipalities and counties formulate targets for their environmental administration here. However, in most cases these targets concern the administration's intended activities instead of the desirable state of the environment. The focus then becomes the resources used in the administration instead of the quality of the environment.

7.3.2 The Implementation

The way local authorities take the initiative on environmental protection depends on the area in question. Their effort also depends on their ability to act. How are they – in practice – able to regulate or affect the behaviour of firms and individuals in their region? The environmental legislation specifies the way authorities can influence and regulate areas where they are in charge, and what kinds of policy instruments they can use in their regulation.

Local authorities find good possibilities for requiring more restrictive demands than the requirements in the law in areas where they have complete competence. This means that municipalities do a great deal in the area of waste management, while the counties do much to protect the groundwater resources. In waste management the municipalities are able to use command and control instruments. This can be done to assure that waste is handled, deposited or recycled in the most secure way. Furthermore, they can charge fees for waste management to cover expenses. In this way municipalities have the ability to handle waste management in their own way, and from the interviews it seems that there have been few conflicts in this area. To limit admissions of discharge water to the sea, lakes or watercourses, the counties and municipalities can use command and control instruments.

Polluting companies need an environmental approval to produce or change production. The Minister for the Environment makes a list containing categories of polluting firms that have to be approved in relation to different kinds of emissions. Furthermore, this list specifies if the county or the municipality has the authority in relation to different firms. In general, the municipalities work out environmental approvals for small and less polluting firms. This kind of regulation is a conditional prohibition against emissions, where the polluting firm gets a permit to pollute up to a specified level. If the municipalities try to set up more restrictive demands on the polluting firms, the outcome of an appeal will often be that the firms just have to keep the standard in the guidelines. As the legal decision then becomes decisive for the environmental targets, it is also difficult in practice to use the local judgement for the actions needed. So, even if these guidelines are seen as minimum standards it is hard for the municipalities to set more restrictive demands for the companies. It should, however, also be mentioned that small municipalities on the other hand find these directions a great help in their administration, for example when the environmental problem is rather complicated.

In addition to the policy instrument, the role of the regulator is important for the environmental policy. If the local authorities are meant to lead a more active environmental policy, their functions or roles towards the population are decisive for the implementation of an environmental policy. As the local authorities in many ways have the legal authority, their function will often be to control firms and consumers. However, many large municipalities find it more important to inform the population about the environment and to encourage cooperation with the population in solving environmental problems. This view also comes from the small municipalities. However, as they often have a stronger budget restriction and at the same time are constrained to make supervision of firms, their resources to do anything, but fulfil the requirements of the law are sometimes limited. The size of the municipalities is therefore decisive for the environmental policy, where big municipalities try to "go further" than the legislative requirements and use other alternative kinds of regulation. For example when they want to improve the environmental performance of the companies by establishing a close dialogue with them and in this way try to promote, e.g. a cleaner production. Such municipalities do therefore have an important part to play in order to handle both local public goods and more general environmental considerations that may perhaps be more national problems. It seems from the survey that the inter-jurisdictional competition in relation to firms is less between big municipalities. This could be because such municipalities are less sensitive to the location of specific firms in their area. The local environmental dumping in relation to firms might therefore increase as the size of the municipalities decreases since they are often more economically dependent on specific firms.

7.3.3 Valuation of Policy Effects

When local authorities have the responsibility, the question is if they have any measures to control or valuate their policy and effort, and if this is done by looking at the effects achieved, that is the state of the local environment after the policy implementation, or the activities that have been fulfilled.

The municipalities and counties have to supervise the polluting firms, ensuring that they keep their environmental approval and some specified standards for emissions to the air, water and soil. If an approval is not retained the municipality or county can apply sanctions on the polluting firm through prohibition or, for example, they may command the firm to take specific actions to meet the authority's demands. The local authorities have the responsibility to control the polluting firms and they can choose how often to supervise them. The environmental agency recommends that the local authorities use one man in their environmental administration per 10,000 inhabitants for the supervision. In practice this means that the central authority sets a minimum standard for the amount of supervision. This indicates in some way that the central authorities evaluate the local environmental policy by the number of supervisions rather than by the state of the local environment. By this, controlling the local authorities limits the possibilities of giving priority to other activities, but the supervision. In this way concentrating the environmental administration on, e.g. the most severe local problems through a selective effort in one specific area is difficult.

Being able to evaluate their effort or effects on the environment some municipalities and counties put up indicators to estimate these effects (Mortensen and Larsen 1994). Sometimes this is done for some natural resources, but more often it concerns the activities or efforts the local authorities consider implementing or have implemented. In this way, municipalities and counties rather evaluate their success as an environmental administration from what they have done and not from the effects this has had on the state of the environment. An evaluation of the environmental action then becomes a matter of money spent and not a matter of the effects on the environmental quality.

7.4 Concluding Remarks

When policy makers want to design an effective environmental policy, it is important to consider if the administrative levels concerned have both the incentives or motivation to solve environmental problems and the ability to do so. These two factors will define the administrative level that will secure an appropriate environmental policy.

In many ways, local authorities do have an important part to play in environmental policy in general. Differences in the state or the evaluation of the environment are arguments for maintaining a differentiated environmental policy. However, if the local administration is going to handle the environmental protection in the most appropriate way, it needs the incentives to do so. This means that local authorities should handle and solve local problems caused by local activities. However, to avoid overall welfare losses, a certain overall coordination is necessary in some areas. Here environmental taxes at national level can be used to include the global and national effects of pollution. So to speak, to make global and national problems local.

Even though local authorities are given the main competence, it seems that some practical barriers exist for the local authorities. The empirical study showed that the local Danish authorities' ability to give priority to special local environmental problems is often limited both because of the appeal system based on national standards and when the municipalities or counties do not have the entire competence. On the other hand, large municipalities and some counties feel that they have a part to play in the environmental policy and do therefore pursue a more active policy in many areas. Furthermore, both municipalities and counties want to pursue a more differentiated environmental policy, for example because of differentiated environmental goals. This indicates the existence of incentives for local action.

The question is whether one achieves better environmental conditions if one transfers power to the decentralized authorities. Evidence from the interviews indicates that if one allocated greater power to large municipalities and municipalities that have no limited resources, this would probably benefit the environment. These municipalities would thereby be in a better position to utilize their local knowledge for a more selective environmental effort, and thereby prioritize their use of resources with regard to solving the most serious environmental problems in the local area. In particular if they are able to diminish the problems with interregional spillovers and local political considerations at the same time. However, by accepting local authorities' ability to lead a differentiated environmental policy one consequence could be that some local authorities lead a less restrictive environmental policy.

References

Basse Ellen Margrethe. *Introduktion til miljøretten.* Copenhagen: GEC Gads Forlag, 1993.

Krarup Signe. *Amter og Kommuner som miljømyndighed.* SØM-publikation no. 13, Copenhagen: AKF Forlaget, 1996.

Miljøministeriet. *Lov om miljøbeskyttelse.* Lov no. 358, 6 June 1991.

Mortensen, Birk Jørgen and Larsen Anders. 'Economic growth, sustainability and the role of local authorities'. In *Challenges to Local Government in European Welfare Development,* Eigil Boll Hansen, ed. Copenhagen: AKF Forlaget.

Naustdalslid, Jon and Hovik Sissel (eds.). *Lokalt miljøvern.* TANO, Norsk institutt for by- og regionforskning, 1994.

Pearce, David W and Turner Kerry R. *Economics of Natural Resources and the Environment.* Harvester Wheatsheaf, 1990.

Oates, Wallace E. and Schwab Robert M. 'The theory of regulatory federalism: the case of environmental management'. In *The Economics of Environmental Regulation,* Wallace Oates, Cheltenham, Elgar, 1996.

Siebert, Horst. *Economics of the Environment.* Berlin: Springer-Verlag, 1992.

Tresh, Richard W. *Public Finance: A Normative Theory.* Business Publications, 1981.

8 SUSTAINABILITY AND CIVIC PARTICIPATION IN ENVIRONMENTAL LOCAL AUDITING. CONTRIBUTIONS FROM THE EXPERIENCE IN CATALONIA

M Àngels Alió, Xavier Kendall and Susana Laredo

Environmental Local Audits (ELAs) and, by extension, territorial environmental audits have in recent years come to be considered adequate tools in order to manage the initiatives around local environmental problems, preferably in urban and industrial areas. However, such a tool for environmental reform has not always lived up to its expectations. This is due to the management and uses which should be modified as well as for the different types of deficiencies of the documents in the ELA itself. Of course, it must be realized that ELAs could have been carried out having in mind a 'green make up' more than a strong will to improve the environmental quality in the context of ecological and social respect. However, all the reasons mentioned above do not invalidate ELAs as very useful tools in order to know the features and implications of the environmental problems at a local scale. ELAs can be useful too when it comes to spreading and indicating the ways of intervention both at the local administration level as well as in the more complex level of everyday life.

The paper we present here makes a start on the basis of such potential change, although we believe that if ELAs are to be really useful then they will have to be carried out in the context of an ecological and social open vision[1]. This means

---

[1] This paper introduces some aspects of an investigation on ELA methods carried out by the Department of Human Geography at the University of Barcelona. The aim of this work was to

accepting the two following working criteria: The first one is to consider it impossible to launch initiatives for environmental change without the maximum support and engagement from the population and, thus, it is necessary that ELAs are carried out in a context of openness to the citizens. The second thought refers to the fact that environmental reform requires the changing of the economic models and some settlement patterns which are at the base of many of the detected problems which we will try to confront.

In order to put this paper into context, it might be convenient to introduce the main features of the experience on ELAs in Catalonia. ELAs appeared in Catalonia at the beginning of the 1990s when the first postgraduate courses with EC funding were carried out in the field of environmental specialization for young graduates[2]. However, it was not until 1993 that the first ELA was started within the specific programme of the Environment Department of the Diputació of Barcelona [3].

At present, the number of ELAs carried out or in the process of elaboration is over fifteen, from which we can distinguish two models of ecoauditing depending on time they were done. The first three audits have in common the stress put on the systematization and separation of problems and the very little consideration for sustainable criteria and civic participation. The second phase, which started in 1995, shows a higher degree of compromise by the municipality with the proposals and conclusions of the ecoaudits. It is also important to note that the Diputació of Barcelona have started to establish a context of previous conditions; among these we can find that the council has to compromise to follow the directives of Local Agenda 21 and it has to join the European Network of Sustainable Cities. It is because of this that the civic participation and sustainable criteria have more relevance in the audit, although there is some reticence coming from technicians and administrative workers.

From the territorial structures point of view it is important to note that the majority of these ecoaudits have been carried out in medium size municipalities, between 20000 and 100000 inhabitants, located mainly in the industrial areas around Barcelona. There are five exceptions: two small municipalities stand out as agricultural-residential municipalities, and three county capitals of the hinterland of the area of Barcelona.

---

provide guidelines for civic organisations interested in promoting these tools in their municipalities (ALIÓ 1997)

[2] Postgraduated Course of Environmental Impacts and EcoAudits, organized by Centre d' Ecologia i Projectes Alternatius (CEPA) and Polytechnic University of Barcelona (1991-1992)

[3] The Diputació of Barcelona is the institution of the administrative support for the municipalities in the province of Barcelona. In Catalonia there are four provinces with these function.

Until the present moment the ecoaudits have generally been carried out by private auditing companies which had great experience and were well known in the field of economic audits and in environmental technology. Such a profile explains why some of the audits have a significant lack in some important aspects, mainly referred to the social dimension of the environmental problems and the role of the relations between local space and global problems. Linked to this type of idea, it is understandable that the civic participation criteria very seldom appear. However, this is being corrected in recent audits in the context of a project which will help to accomplish the different ways of participation. Recently, some civic and ecologist entities and university scholars expressed to the Diputació of Barcelona their views and their proposals for future ecoaudits. Such pressure has sparked the reinforcement of more civic aspects and the prevention of environmental problems, which has helped to introduce the discussion about the territorial models at a local scale in the administration.

8.1 The citizenship involvement

The importance of the citizens' involvement as a requirement for the success of environmental policies has been stressed many times. This has been done so often that it is now a common issue in technical introductions and proposals for intervention. However, we are not discussing the level of truth in such statements, but we want to draw attention to its advantages if we are to endow them with a real content and achieve better results when we have indulge in self-criticism, which is what environmental auditing is about.

8.1.1 The democratic and participative component in the environmental urban reform

For three decades citizenry has started to have an important role in the environmental discussion and in the way of intervention in and assessment of the impact the model of economic growth and urbanization which has headed the organization of the territory. Although this was first restricted to ecologists and volunteers, the preoccupation with such issues has increased following both the increase of the environmental problems and their social consequences as well as the failure of the traditional political structure of involvement. So we could say that the dynamic environmental implication looks similar to other groups which have organized themselves around other problems, social, ethnic or cultural. These issues are usually left out as political priorities, but when they are introduced through citizen mobilization, the same problems become political issues (Alió and Bru 1992).

So, our proposal concerning ELAs is nothing other than a way of trying to reinforce the possibilities of local citizen involvement in a context in which the administrations are experiencing a crisis of solutions and in which alternative proposals are beginning to emerge as acceptable solutions. For instance, this is happening with the issue of waste minimization and pollution prevention.

Once we plan the matter in such a way we are going away from a basically technical idea of ELAs; we are heading towards a plan in which documents are elaborated in a process which is coordinated by the auditing team. But associations, local people and citizens have a say and we have to admit their active role as a being involved in a series of interventions and proposals affecting them directly and about which there is a wide uncertainty (Funtowicz 1995). This point justifies the political consensus for citizen involvement, and it is stronger when we consider the deficiencies and uncertainties of the scientific working methods about many of the environmental problems. Populations affected by these problems can contribute with precise and specific information of different kinds (Eden 1996).

It is commonly accepted, though, that the aims of the environmental reform also require a whole series of changes in everyday behaviour, relations and social values. We then find that citizen involvement, which we have considered up to now just from the angle of the new relations in politics, is connected with the fact that these changes lead to the emergence of an alternative system of ethical and ecological values (Garí 1995).

This public and individualized dimension of engagement and citizen responsibility in new forms of relationship with nature puts us face to face with the fact that the changes demanded are within a more complex context of transformation which also affects the productive and management system (Alió, Cervantes and Kendall 1994). Besides the aims of ELAs as tools for 'ecological' change in municipal management, we cannot forget that ELAs, because of their specific character, are also an ideal tool for studying and publicly discussing the engagement of the local economy in the formation of related environmental problems, and thus reinforce the strategies of the reform in exchanges and industrial systems.

Probably, such a view will enable us to put ELAs and the other tools of environmental reform into a much wider historical context, which has its roots in the past and is projected towards a future in which the issue of involvement and engagement becomes increasingly decisive. It is obvious, though, that by now we have to work on the hypothesis that it is very likely that such a type of more compromising ELAs, which are more scientific and effective, will be generally accepted.

It was no accident that when carrying out this project we were thinking about municipalities where population and local administration were already sensitive when it comes to search for ecologically acceptable solutions.

However, there are some environmental audits and environmental reports which demonstrate such types of planning. It is the case for example of the development of the environmental urban reform in certain Scandinavian countries or the example of the tradition of the USA citizen consultations. Thus, it is interesting to remember the contributions of some ecological organizations in the methodology matters of making ELAs, and in the analysis of the impact of management and economic activities (FofE 1992). These contributions go, without doubt, towards the reinforcement of the public dimension of these types of documents.

8.1.2 Citizen involvement during the creation and the follow-up processes

In relation to the development of an open and active model of local environmental audits which would help us to consolidate the citizen involvement and engagement practices affecting environmental quality, we have set up a working project programme in which we have limited the different responsibilities and competences of the three parts in ELAs: the auditing team, the local authority and the civil entities and general public. We are about to present the different ways of civic involvement which are considered to be suitable.

8.1.2.1 Citizen involvement during the process of creating the ELAs: The different types of involvement. In the same way as we can identify important differences between the several stages in the making of the audits in consequence of the succession of several methodological sequences. The types of involvement and citizen engagement also change. We could say these evolve from some first bridging initiatives between the population and the team towards, in the final stages, the making of a group of measures which require important levels of compromise and engagement from the citizens.

Generally speaking, such evolution almost coincides with the four main stages of the audit, which are: (1) The preliminary stage: for drafting the guide-lines of the work which will rule the work and the writing of the documents. (2) The descriptive stage, which will enable us to know the state of the environment in the area and its problems; followed by (3) the stage of diagnosis: which implies the studying of the problems we have found in the territory and the assessment of their social incidence and the degrading of the environment locally and globally. (4) The stage in which the proposals are established: such proposals provide the most suitable solution for each of the problems (Table 8.1).

Table 8.1 Mechanims of Civic Participation during the auditing process

1. PRELIMINARY STUDIES	- Informative workshop about ELA and its objectives in the municipality. - First approach about the possible collaboration according to type of association. - Establishing different collaboration mechanisims during the audit. - Discussion workshop of the hypothesis of the ELA (sustainability, assessment criteria, etc).
2. ANALYSIS	- Meeting in order to search and consultation of specific information. - Making the inquiries. - Helping with field work. - Discussion of the work document by informative workshop or other types of meeting.
3. DIAGNOSIS	- Global sessions with the first results about environmental problems. - Specific workshop with the most affected population. - Discuss the working document about the results of the assessment.
4. PROPOSALS	- Information about advance of proposals and its objectives. - Discussion about establishing specific proposals. - Public exposition of the report, diagnosis and proposals. - Allegations, Proposals, Claiming.

Therefore, in the preliminary stage the ways of involvement will be important. Such ways will be considered to establish the first meetings between the population and the auditing team. These meetings can be organized by local entities, the Council or the team. In any case, though, it is necessary from the beginning that the auditing team sets up a working plan in order to establish the meetings with the citizenship and the different civic and economic entities. This will help us to be aware of the chance to establish ways of cooperation in the development of specific supporting tasks. Another important aim of these meetings is to let everybody know the hypothesis used throughout the work and which is expected to provide the starting points of a public debate about the local possibilities of change towards sustainability. It will also provide a knowledge of the existing relations between social-economical attitudes and environmental problems and also about the role the citizens have to play in the process of environmental reform.

In relation to the following stages of analysis and diagnosis, it is foreseen that at these stages the meetings with local entities will continue, although at this stage, on the basis of the search for information, the meeting will concern general opinion, the problems and situation of the environment.

We foresee, though, that a number of such data can be obtained through interviews and inquiries which, directed by the auditing team, would be interesting if they were carried out by members of the entities. We refer here to workshop schools, local ecologists and others who could do a great task.

It is also foreseen that the expanse of the ideas and criteria which go with the creation of the ELA could find a diffusion channel based on of the thorough interviews carried out with important members of the community. The relation with them should be seen as a way to self analyze and begin future action.

As for the assessment of the problems, it must be said that in the stage of diagnosis some of the inquiries should lead us to know the perception and social value of the problems detected. This is because it is considered that the views of the population constitute a legitimate assessment criterion and it is complemented with other criteria considered to be more effective. These last criteria are the impact analysis and the legal standards of environmental quality.

We have to consider that it is foreseen that by the end of the second and third stages, a working document will be produced containing the most important results which should be well published and the subject of discussions in the meetings between the team and the entities. As we see it, it is not about introducing new issues now but of spreading widely a very usual practice in environmental auditing and which would be useful in this case, to deepen its self-critical and engaging function among the different population and different types of collectives.

As for the fourth stage, it must be pointed out that two different types of involvement can be distinguished. First, we have to carry on with the information tasks, although these are aimed now at explaining the intervention and monitoring proposals. What we try to achieve with this is the maximum support for several proposals which could be difficult to accept for certain people. We also try to collect ideas and suggestions for matters not put through before. In short, we try to sum up the technical stage making proposals with the maximum spread and consensus in order to achieve the maximum depth in the final drafting of the proposals and in the implementation programmes for both facts and changes considered conflictive for problems concerning the economic interests at stake. That is, for example, the case of the industries with serious impact on both local and global scale and, on the other hand, the problems directly linked with day-to-day behaviour.

However, the final stage, when decisions about the proposals are made, is no more than a stage when, even though it has to coincide with the structures of administrative procedure, the procedure of direct participation cannot be forgotten.

For this reason, our idea brings the fact that both the population on one hand and the Council on the other (Guiford 1991), have to compromise formally with the ELA and its objectives and proposals. So, before summing up the definitive documents, there will be a stage of public and open exposition followed by a period of allegations to study by the team and by the Council. This period can provide new meetings, helping to clarify criteria and positions in the face of the initiatives which will be approved. It must be understood that, even though the public exposition period is quite clear according to Spanish regulations, it is hardly ever used in practice, with the exception of expropriation cases. It would be positive, then, that ELA would prove useful in order to go forward in its experimentation as an instrument to deepen and revise the proposals.

Once allegations are resolved, it is finally foreseen that the final document would be approved by all the Council officers and therefore an institutional compromise allowing continuity to the process is established.

8.1.2.2 Citizen involvement in the follow-up of the proposals. The role of the district and local entities. As we mentioned at the beginning of this paper, the auditing process is not only an isolated and single process of environmental reform. It is part of a much wider historical process. This is why the measure for the monitoring, of the environmental situation, the accomplishment and adequacy of the adopted measures have to be incorporated in the proposals of the ELA. This monitoring coordinated by the administration and planned under the open focus, has to foresee the addition of mechanisms of civic involvement which belong, in fact, to the same processes of engagement and civic responsibility which had to be launched during the audit.

Resulting from this idea we proposed the possibility to use the district as a unity of monitoring. Districts reflect the city dynamics. This is because the district is the area where, through day-to-day life, the problems and expectations of the population are shown. In the same way, it is also the area in which neighbours share responsibilities or react to problematic situations. It is also where we can see more clearly the role of the administration when it comes to solving/generating problems affecting people's lives. On the other hand, the gathering of information on the different urban variables (water, land, health, etc.) is very well particularized in such a unity. Thus, it is easier to understand the differences inside the urban form in order to improve the intervention strategies and favour citizen engagement in the project for environmental reform based on specific actions and proposals for the districts. It would be interesting to include in the audit monitoring team a number of commissions formed by representatives of neighbourhood associations or ecological groups which are sensitive to specific environmental problems.

8.2 Sustainability as a reference

As we mentioned at the beginning, the concept of sustainability constitutes a frequent reference in environmental policies, although very often its use is more in terms of a general heading than with a real and specific will. However, it is clear that this fact does not have to imply losing sight of its potential as a guiding idea underlying many of the initiatives which should be implemented in order to change the social and economical processes which are the origin of many environmental problems, both at the local and the global scale.

This is the reason which made us pay special attention in our project to raising the question of what could be derived when using such a concept in methodological aspects and when formulating proposals. This second part will concentrate on this issue and especially in the social aspects of the implementation which has important implications when it comes to face the problems of urban areas.

8.2.1 Sustainability as hypothesis

The first aspect which has to be considered is referred to the potential of the concept of sustainability as a guiding idea for initiatives and reports on environmental problems and their solutions. We are all aware, though, that introducing the concept of sustainability as an hypothesis can seem obvious if we do not specifically put it in a place within a context of reality in which unsustainable tendencies and structures are dominant. So, to state clearly the idea of sustainability means a number of risks and conflicts both from the inside and the outside the perspective of the audits. But, if these risks and conflicts are solved, the solution can rebound favourably in a better use and more precision when analysing and making the proposals.

One of the key factors of these conflicts lies in the political and social dimensions of the concept. So, in terms of the population, it is obvious that the concept of sustainability is not the same for everybody. ELAs should help to increase the knowledge of the basic economic notions of sustainability and its ecological and social dimensions. It should also help to increase our knowledge of the local and global problems linked to the present ecological unsustainability of many of the existing relations with nature. Therefore, it is important to create a local discussion about the social and ecological implications of some of the less sustainable everyday practices. The debate should include the dominant economic model in the municipality and in the county because it is necessary that audits face these questions openly. In this way, ELAs start up a public debate of an issue which has many consequences on environmental management and can at the same time develop many of the proposals for the audit.

As for the role of the concept in the smaller context of the teamwork, it is clear that the concept of sustainabilty has a very outstanding place for several reasons. A very important one has to do with the fact that one of the aims of ELAs is to show the way and the potential of local sustainability. Sustainability, or the lack of it, is then a theoretical and methodological idea which cannot be faced superficially. Even though, on all stages of the elaboration of an environmental audit, sustainability has to be a reference. We would like to stress the role of sustainability when it comes to assessing environmental problems and activities which have an impact within the municipality. From this perspective, the assessment can be carried out from a multiple and open angle which includes the influence of several social and economical factors as well as the traditional legal criteria and physical and chemical measurements.

We could mention as an example the assessment of the problems related to the future consequences and risks of the infrastructures of waste management, with an important capacity of consolidation of the present economic growth and its territorial impacts. Another issue included in this group is transportation; its environmental assessment very seldom points at the impacts caused by its role of enforcement of the unsustainable present model of territorial organization. The assessments do not point either at global warming and the depletion of non-renewable energy.

Coming back to the people's involvement in the discussion about sustainability, it must be said that in recent years we have witnessed in Catalonia that the appearance of sustainability has started as a result of citizen demonstrations against the two types of infrastructures previously mentioned. There were especially important the mobilizations against the Plan for Hazardous and Toxic Industrial Waste (1991) (Alió 1991) and (Demay and Frost 1992) carried out by the inland counties, mainly rural, and which were affected by a programme planning to build landfill sites and incinerating plants. Eventually, the programme had to be withdrawn. As for transportation, we can say we are presently in a phase of protests against the construction of the fourth belt around the metropolitan area of Barcelona. It is feared that such developments will affect some areas which still have some environmental quality.

8.2.2 Developing the concept of sustainability

Probably, one of the reasons for the acceptance of the concept of sustainability lies on its relational character. Such an advantage, which helps to connect nature with economy, needs to be used accurately, requiring an effort to get closer to the reality of the study. It is clear that such an effort to concretize the conditions of sustainability means the integration of the social and human dimensions of the phenomenon as well as the territorial features of the area and its global connection.

8.2.2.1 The Social Dimension of Sustainability. One of the commonly accepted conditions of the model of sustainability is the one stating that sustainability cannot be achieved if it means worsening the breaks and inequities in society. We would add that the social dimension of sustainability is the ethical and political component of the sustainable use of nature. Therefore, the concept carries a number of values that make it assimilable in the context of new environmental policies.

Once the issue is defined in such way, we believe that environmental audits should take into consideration the social problems related with environmental problems. However, this type of method is presently far from the conventional practices of ELAs and from environmental policies in general.

This fact cannot make us forget that the emergence of environmental audits has arrived at a moment in which urban policies are conditioned by the precedence of the idea over urban transformations and reforms. These latter defined basically from the possibilities of intervention coming from the urbanism and the economy. It would be desirable, then, on top of the concept of sustainability, that really integrating projects of the different branches of urban and territorial problems could be impelled.

As for our project, it must be said we tried to analyze the social problems from the different chapters of our research[4] and we especially focused on health due to its relation with phenomena and processes such as pollution, work, housing and food, among others.

---

[4]Specifically we have considered the following subjects: Urban areas, Natural environment, Air pollution, Water pollution, Land pollution, Noise, Energy, Waste, Transport and Health.

Even though we frequently find that the references to the social dimension of sustainability seem like a compensation for the acceptance of ecological unsustainability of the city and urban areas in general. So in this sense, the point would seem to begin with the hypothesis of the intrinsic social sustainability of the city as a result of its role in the strengthening of the mechanisms generating jobs, concentration of services and its role in the development of public and cultural systems.

Such a question, which does not mean not trying to improve environmental quality, differs from the ones arguing that even given its clear social advantages, the urban phenomenon belongs to a single historical process on a global scale which presently means the strengthening of social inequities in cities and urban areas in the First and Third World.

8.2.2.2 Territorial Dimension of Sustainability: Conflicts due to the relation between urbanization and territory. Furthermore, on the ecological and social dimension of sustainability, it has to be considered that such a dimension is local, although it is formed by a number of situations which go beyond the territorial specifics and lead us to the global structures. This fact makes us consider the two basic structures of the global scale – the ecosystemic and the economic – and the ways in which these are particularized on the local scale.

To focus properly on the relation formed between the specific character of local problems and the global context of the phenomenon we have taken the concept of territory used by geography. From this concept we are especially interested in pointing out the inputs which insist on the social dimension of the relations with nature inside the regional areas. Such view, can be applied on the issue we are analyzing because of two reasons: First because it is very concerned with the social, economic and geopolitical reasons for the spatial boundaries (frontiers, municipal limits and so on). Second, because it enables us to enter the issue of social relations which fit the use of nature in the areas of study. If we take into account that the municipalities are the smallest territorial unit in which the double level of relations between work and land is concretized (Raffestin 1982) we will realize that such a concept enables us to analyze the specific and particular social processes which take place at a local scale but influence the global structures.

However, now it is not about entering the issue of the connections between local and global but to analyzing the issue of the unsustainable spatial structures which are leading the relations with nature in countries like ours. It is also about how a process of civic and social conflict and positioning has been developed around such structures and its alternatives. From these, the environment is an important factor, even though it has not had the appropiate attention until very recent years.

(a) Urban structures of territorial organization and the role of the municipalities. The countries of what is known as the First World are mainly urban, which implies that the majority of environmental interventions are carried out by organized structures based on the primacy of the urban system based on the presumption of unlimited economic growth. This all amounts to the paradigmatic model of expansion and maturing of the urban regions; from this view it is believed that the perspective of the future go through the restructuring of the central spaces and the urban territorial expansion along with the intensification of mobility, decentralized industrialization and the undefined expansion of the urban systems. Therefore, this model does not contemplate any dramatic problems in the processes which are pushing the model forward and which at the same time are generating the environmental problems of our day. The model also relegates places to a subordinated function of the expansion processes and the maintenance of the system. This idea of the municipality as an undefined space and, therefore, exchangeable, depending on the needs or conditions of every moment, means the loss of the potential of the small local ecosystems as well as its loss of value of social conflicts provoked by the loss of expectations of parts of the population in terms of economical potential or use of their own space.

The domain of this model is opposed with the changes put through from the model of sustainability. This adds something to the conflicts which have been going on through history caused by the differences among social groups due to the expectations and interests linked with territory and environment.

(b) Local society facing the challenge of sustainability. Such differences lead us to consider the conflicts which might appear between the local community because the implementation of solutions destined to promote the model of sustainability.

However, this is not the place to go into a detailed description of these inputs which help to make clear the environmental positioning of the different groups. But we would like to show in a simplified way the tendencies which characterize the local debate on environment. The first one joins all the groups which are impelled by the eruption of the environmental variable in local policies. This is a wide and diversified tendency, supported on the current ecological wave, which give a new vision of the urban movements organized around the use values criterion. In addition, this tendency is impelling the transformation of the democratic system in present society (Reichman and Fernández Buey 1994). The second is centered on the sectors which hold economic relations with the environment. Nowadays, and as a consequence of the recent transformation of capitalism, such types of relation mean very frequently the appearance of new conflicts caused by the different interests and strategies within the sector (Watts 1989). On top of this confrontation, we have to consider those who believe that the betterment of environmental quality is the indispensable condition for the reproduction of the economic system.

At the same time, it is worth noting that from the point of view of sustainability, different autors have stressed opposite positions on the notions that sustainability exists (Escobar 1990). In the search for a balance, some administration try to place themselves between those opposite ideas with different results.

In order to follow our explanations it is interesting to focus now on the collectivities and entities directly involved in the defence of environmental quality, either from purely conservationist positions or clearly ecologists. It is among these that the local knowledge can be found which has to be taken into consideration when it comes to formulating the proposals on the development of the model of sustainability (EDEN 1996).

The present tendency of growing degradation of environmental quality as well as its expansion to be a world wide problem has given those groups a key role. They have strengthened their criticism and opposition to the new social urban and territorial movements to the point that environmental defence has become one of the key issues in popular demands.

However, such a tendency does not leave behind the chance to meet some conflicts which can develop very differently depending on the positioning of each group, the mobilization of the citizens, and so on.

Even though the following examples are not actually related ELAs, we would like to mention some conflicts in Catalonia where the environmental debate was referred to the local environmental model. One of them is the conflict already mentioned for Plan on Hazardous and Toxic Industrial Waste, which led to the association between the ecologist movement and the farmers in a region based on the production of wine. The farmers were interested in protecting their interests against industrial plants which were seen as the first indicator of an invasion of polluting industrial activities coming from the urban centres. Another example, in this case demonstrating the agreement between the popular classes and the ecologists, can be found in the movement for waste minimization and against the incineration plants. From another point of view we can also mention examples of environmental mobilizations which were not successful because of the alliance between economic interests and the working class. The case of the installation of a paper factory in the area of the protected water supply of

Mediona shows very well how the strategies towards sustainability can be set aside if the social forces are not pursuing the same objective.

(c) ELAs as tools for opening the way towards sustainability. A consequence of all we have been mentioning is that the objective of sustainability and the formulation of proposals on an ELA can clash with interests and strategies which differ in scale. In a very simple way we could distinguish two opposed ideas, on one hand the ecologist' local groups which are interested in using ELAs to promote preventive and restorative policies and on the other hand certain economic and political groups, generally well connected with the socioeconomic global structures, which foresee new economic functions in the context of the present model of growth.

At this point it is important to recall what was said before when analyzing the civic participation and its role as an active factor in the process of the implementation of environmental policies in their local aspects. So, maybe it could be said that even though it is foreseen that all social groups are interested in improving the environmental quality of their city or municipality, it could very well be that there are differences between them in terms of the assessment of the problems and the formulation of the proposals for their solution.

It is then foreseen that the performance of a Local Audit will provoke discussions and problems linked with the different models of local economy that different groups and collectivities have. However, a Local Audit is not a Strategic Plan, not even a Local Development Plan. It would be useful, then, that the auditing team had very well prepared arguments. The perspective of sustainability makes everybody replan the problems of local economy and the relation between the municipality and the territorial structures. The ELA would also help to start a debate about the environmental problems of the economic base and the everyday behaviour of the citizens – for instance transport and consumption. Another important use of that debate is to start up proposals and local alternatives which, in another way, would not have a place in the traditional local policies. Such proposals and alternatives could be developed in a future programme for the reduction of waste, the ecological, industrial and agricultural change, ecoplanning and so on.

To finish off this point, it would be useful to add two more issues which are important from the perspective of the relations between locality and territory. The first one is that the model of sustainability includes an active idea of localities which are interlinked in a network of balanced and mutual relations among all municipalities. With this aim, ELA should help municipalities to find criteria and alternative ideas for the programmes for urban expansion into the territory. Finally, the second issue has to do with the technological dimensions of the environmental reform. These could be thought as dominant criteria and proposals instead of the substituting processes and the dramatic changes the model of sustainability requires. Put it in a different way, it is not acceptable, for instance, that under the presumption of civic participation the proposals would end up promoting a process of spillover of pollution or waste to other villages instead of helping to find solutions destined to prevent the problems of pollution at its origin

8.2.3 Sustainable proposals of intervention for management of urban areas.

To initiate strategies for urban sustainability means having to intervene in many frontlines or areas connected with each other. Such complementarity is shown when, in the process of making ELAs, the issue of the proposals is faced and the deep connection between society and nature becomes obvious. In the field of Council management, such issues are enlightened by the possibilities to reinterpretate the conventional instruments of management – planning, social affairs, populations, etc.– in the light of their relations and their capacity to generate situations capable of starting effects of change and improvement of the environment. Even though it could seem that this idea is opposed to the issues considered in environmental audits where the areas of study and intervention are faced separately.[5] We think it could be interesting to point out this fact in terms of the demands for change of the reports and the management systems used in the solution of environmental problems.

As an example of all this, we have concentrated in this section chapter on the issue of urban planning and its role as instrument of environmental reform. It must be said that even though in some countries urban planning is situated within the context of the systems of integrated environmental intervention – as it happens in Denmark – the tradition in Mediterranean countries is of a very scattered practice basically away from ecological problems and led by criteria of growth.

This attempt to introduce planning as an instrument for a sustainable urban environmental reform has been carried out based on three points: ecological, social and economic. For each one we will try to show some of the most important connections with the rest of the environmental audit as well as its interest from the perspective of the citizen involvement and the public debate around sustainability.

---

[5]According to the scheme of the Green Book.

8.2.3.1 Ecoplanning. Even though the origins of contemporary urbanism are influenced by strong environmental concerns, it is not until the end of the century when we can find ecoplanning inspiring many of the present proposals of environmental change for the cities. It could very well be that one of the most interesting ideas lies in the individualized set up of ecoplanning. This means that each city or municipality has some special features derived from the differences in the urban society and, on the other hand, from its position in the ecosystem which establishes a number of potential and conditioning features. These features must be respected when taking urban decisions (Deelstra 1992). Such positioning, which takes us far from the functionalism argument of conventional planning, puts us closer to a different vision of the expectations and initiatives to conduct in cities and in which environmental audits are related with issues such as biodiversity, hydrographic systems and climate among other physical variables.

There is a second very interesting idea underlying ecoplanning. We mean the issue concerning the different fields of work which, as kinds of subsystems of global ecoplanning, establish the different types of intervention. These are: ecosystem, water, mobility, energy, waste, architecture and planning. This last in the sense of ecological urban planning with important consequences in the location of the uses, the densities, the green between others. All of them are integrated in a new vision of urban management, not dependent on corporate or economic interests around the built environment or location of activities but which is decided in the frame of ecosystem differences, conditions and possibilities.

As an example of this idea it is worth mentioning that in the proposals for our project, in the transportation chapter we have been forced to incorporate proposals intervention of a very urban nature. The same has happened in the chapters on waste, land pollution and also in open spaces with its obvious connections with the proposals for natural environment and energy.

To approach the social reality of the city, with its differences and conflicts, and change it in a factor of an even-handed urban management with solidarity is one of the political challenges of the present day. From within ELAs, this issue is placed within the framework of the social dimension of sustainability. It also has its place through the empirical evidence of the problems studied and the contrasting and public debate generated on the basis of the different ways of citizen involvement which have been mentioned earlier.

As for the empirical aspects of the work and the formulation of proposals, we have followed two main lines in our project. On one hand, we have stressed the use of instruments which have enabled us to collect and represent cartographically some problems according to their distribution in residential areas. Some of the most important issues are the problem of sewage disposal, the location of landfill sites, either controlled or uncontrolled, and the incinerating plants as well as the location of industrial centres creating pollution. In the line mentioned earlier, when we were referring to the chapter on health, it must be stressed that at this point a programme was elaborated to collect data and put it in graphs. This was carried out in order to assess the existing relation between these and other problems and the inhabitants' health.

The second line of work is centered in the district. We already referred to it when we were discussing the instruments of citizen involvement in the follow up after the audit. The reason for insisting on the district comes from the fact that the spatial concretion of the problems is collected in a number of items of information, telling us the different location tendencies of many of the environmental problems affecting population can be approached with the global view of the district or the smaller residential areas which form the urban structure. Therefore, in the report phase and the proposals on urban matters, a specific treatment for each of the residential areas is previewed. Such treatment will gather all the problems affecting the residential areas and the proposals to solve them and later it will be possible to monitor their solution.

8.2.3.2 Sustainability and urban economic base. However, it is not possible to impel strategies towards sustainability in a context of a regional process which, on top of the centrifugal tendency from the central areas towards the metropolitan outskirts, spreads the adoption of a non-ecological model of economic and urbanizing growth.

This fact leads us to consider two questions. The first one is referred to the local and global connection to which we have already mentioned and the role of ELAs as a leading indicator of local potentials of sustainability. The second question takes us again to citizen involvement and its role in the debate in local expectations towards sustainability.

As for the first question, we could say we are presented with the difficulties in facing impacts and changes generated by the advance of the metropolitan texture and the need to preserve and impel a number of expectations. Such expectations had not deserved enough attention up to now, but from this moment can obtain a privileged social and economic function when the hypothesis of sustainability is considered.

The same could be said for the central urban areas which are changing, although in different ways. These areas have to find a solution to a number of problems too. Among these, as in the case of Barcelona, we could mention the shrinking tendency of public transport (Casals 1995).

However, it is clear that along with these local problems, we have also to consider the problems derived from the central function of these cities. This is impelling the present model of non-sustainable territorial organization. At this moment, one of the challenges for environmental policies and for groups involved consists precisely of finding strategies for territorial sustainability which will enable us to respect many of the advantages for the society of these cities.

Consequently, ELAs should introduce proposals for the changing and the endowment of a local economic base more respectful of the environment and which, along with the urban, conservation and demographic aspects – such as the one referring to dwellings, ecosystems, waste minimization or even when necessary to stop the demographic growth – should also include indications about the behaviour of the agents and the local economic base. In fact, we have already mentioned the difficulties which might provoke these changes. Even if it means taking the risk of simplyfing the issue, we would like to suggest that the approach we have taken to sustainability means taking into account the idea of time when it comes to analyze the proposals. So we should differentiate, on one hand, proposals meant to impel dramatic changes and which cannot be introduced without either a time prevision of the initiatives and a general agreement within the municipality.

Table 8.2 Some proposals on ecoplanning

1. Structural basic aspects	- Modify the urban prevision towards a slow demographic growth which will favour a stabilization of the population. - Enlarge the preserved areas in the context of the change of the urban qualifications and the uses of the land. - Include conservation and sustainability in sectoral local planning.
2. Local specificity	- Protection of the local specificity: . local ecosystems and the promotion of their resources . historical heritage and quarters . diversity of the economic function -Promotion of the ecological district entities
3. Urban planning	-Substitution and modification of the architectural and urban design within the ecological context concerning: . enhance the green areas within the city . enhance walking trips and the use of public transport . enhance new urban demands created by the separate collection of waste, the saving and renewable energy - Modification of the by-laws in order to achieve the last three objectives.

Table 8.3 Proposals on local sustainable development

1. Local non-sustainable activities	- Follow-up on their impacts and potential hazards. - Time table for the future discussion
2. Ecological change of the economic companies	- Assessment of the role of the local recovery and recycling companies. - Assessment of the local results in campaigns to change industrial processes. - Develop the role of the local public corporation compromise on the adoption of the ecological changes in the industrial and commercial processes.
3. Local development initiatives	- Study of the local political development iniciatives. - Assessment of their impacts and follow-up. - Promotion of new sustainable initiatives (EcoPark, etc).
4. Local sustainable potentialities	- Human resources: craft, culture, information, education, etc. - Economic activities: ecological agriculture, rural, tourism, clean technologies. -Physical environment: urban reform, old city, landscape, wildlife, physical resources, etc.
5. Alternative energies	- Study of the local initiatives on alternative energies (private and public). -Study of physical local potential to implement solar energy and wind power. -Promote the use of alternative energies in public buildings. - Promote the local public corporation in the creation of local companies for renewable energies.

These proposals mean different types of debate and information diffusion among the citizenship and the economic agents. This debate is destined to reveal the interest and obstacles which can be found when the proposals are be implemented. It is eventually about proposals which mean the acceptance of sustainability as a model which is projected towards the future and from which we are just beginning to see new ways. On the other hand, we have the initiatives which can be taken in the short term. Among these, and with direct implications on economy, we are interested in stressing a wide variety of proposals with a great capacity to introduce changes in the many municipal relations. We are referring specifically to the ones related with (a) urban planning, with economic implications through the urban structure, the housing market, exchange and growth expectations (Table 8.2), and (b) the ones directly affecting the local economic base (Table 8.3).

These second type of proposals could be formulated in five areas: (A) Local economic sectors which are not sustainable and which are difficult to change on the short and medium term. (B) The recent initiatives of local development promoted from municipalities and the assessment of the expectations of sustainability. (C) The new ways of local development based on the promotion of human resources – crafts and different forms of services and cultural development.– or the natural environment, ecological agriculture, quality of life, leisure economy, etc. (D) Assessment of the local or regional potential of alternative energies and indication of initiatives to develop by local institutions towards such aim. (E) Possibilities of ecological transformation of the local economic base, showing the available instruments such as initiatives for the change of polluting industrial processes, etc.

Finally, we could go back to the issue of citizen involvement and the need, when considering local sustainability, that wide segments of society will not be excluded. Such need comes from the fact that local sustainability means considering a number of facts and variables which go beyond the technical discussion connecting with the social and political dimension of the ideological and cultural ideas of people about their villages and cities. Only for this reason we should insist on the need that people have to take part in the decision around these questions. ELAs could be one of the necessary ways to reveal and consolidate the public debate about local economic models and their global, social and ecological consequences.

REFERENCES

Alió MA (Coord.): Cap a la reforma ambiental urbana: Aportacions des de la metodologia de les ecoauditories municipals. Barcelona, Ed. de la Universitat de Barcelona, 1997, 170 pp.

Alió MA: El discurso ambiental en la gestión urbana: Del urbanismo ecológico a las ecoauditorias municipales, 'Revista de Geografia', Universitat de Barcelona, 1995, vol. 21–35.

Alió MA: Gestion du milieu et organisation du territorire. Controle et Traitement des Dechets Industriels en Catalogne. 'Villes et Territoires', Toulouse Le Mirail, 1992.

Alió MA, Cervantes A, Kendall X,: La difusión del modelo de ecoplaneamiento en el marco de las iniciativas para la calidad ambiental en Catalunya. 'La industria en la planificación urbana. V Jornadas de geografia industrial', Girona, 1994, 25–33.

Alió MA, Bru J,: L'esquerda ecològica: Residus industrials i geografía humana. 'Documents d'anàlisi geogràfica' Universitat Autònoma de Barcelona, 1992, No 19–20, 11–32.

Alió MA, Kendall X, Laredo S: Sustainability and civic participation in Environmental Local Auditing. Proceeding of the Ecology, Society and Economy. In pursuit of sustainable development. International Conference of the European branch os the International Society for the Ecological Economics; Saint Quintin en Yvelines, V-1996.

Casals V: Transport, territori i medi ambient. Barcelona, Secretaria de Política Sectorial i Medi Ambient de la CONC, 1995.

Centre de Ecologia i Projectes Alternatius: Ecoauditories Municipals. Document de Treball . 'Seminari sobre la reforma ambiental urbana', Barcelona, VII-1995.

Deelstra T: The city and its resources: Thoughts on sustainable urban development and the quality of life. 'EcoCiutat', Barcelona, 1995.

Demay A, Frost S: The controversy concercing industrial waste in Catalonia. 'Environmental Resources Unit', University of Salford, 1992.

Eden S: Public Participation in environmental policy: considering scientif, counter-scientific and non-scientific contributions. 'Public Understand and Science', 5, 1996, 183-204.

Environmental Auditing. 'Environmental Practice in Local Government', 1992.

Escobar, A.: El desarrollo sostenible. Diálogo de discursos. 'Ecología Política', 9, 1990, 7-25.

Friends of the Earth: Environmental audits of Local authorities: Therms of reference

London, Friends of the Heart, 1992.

Garí JA: Propuesta de una ética global. 'Congrés internacional tecnología, desenvolupament sostenible i desequilibris', Terrassa, 1995.

Gyford J: Citizens, consumers and councils. Local government and the public. London, Macmillan, 1991.

A guide to the Eco-Management and Audit Scheme for the UK Local Governement. London, HMSO, 1993.

Funtowicz S: Problemas ambientales complejos y la ciencia postnormal Proceedings of the Congrés internacional tecnología, desenvolupament sostenible i desequilibris, Terrassa, 1995 desember; Barcelona, Icaria Ed., 1997.

Raffestin Cl: Remarques sur les notions d'espace, de territoire et de territorialité. 'Espaces et Sociétés', 41, 1982.

Reichman F, Fernandez Buey F: Redes que dan libertad. Introducción a los movimientos sociales contemporáneos. Barcelona, Ed. Kairós, 1984.

Watts M: La geografía y la lucha en defensa del medio ambiente: la integración de la ecología humana y la economía política o ¿Marx era rojo o verde?. 'Boletín de la Asociación de Geógrafos Españoles', 9, 1989,109-126.

9 LOCAL SUSTAINABLE DEVELOPMENT: *HOW CAN EQUITY ISSUES BE EXAMINED AT THE LOCAL LEVEL ?*[1]

Jane Corbett and Heather Voisey

Introduction

Both equity and action at the local level are seen as critical to most definitions of sustainable development. This is perhaps best captured in the phrase 'community based sustainable development'. But the term community is often used to imply action at the local level, without either community or local being clearly defined. Furthermore the term community is often used as if communities were homogenous in economic and political terms, suggesting that the only equity problems to be resolved will be differences between communities. This neglects consideration of how significant equity, specifically the existence of economic and political inequality within communities, will be to the ways in which the challenges of sustainable development are addressed at the local level.

Many of these problems existed before the term sustainable development had been coined and have been much discussed in other contexts, for example in analysis of the relationship between economic development and equity in low income countries. But the challenges of sustainability intensify the need for effective interdisciplinary research and action which links the insights and analytical methods of the physical and social sciences and increases understanding of how environmental and social change interact. Clarity about the

[1] The authors would like to thank Neil Adger for his helpful comments during the preparation of the paper.

use of the terms equity, local and sustainability will be critical and so will the development of new methods of observing and analyzing how equity issues are handled at the local level in the context of sustainable development.

This chapter takes as its starting point the assumption that there will be differential impacts within communities of any external or internal changes involving the regulation or use of natural resources. It resulted from collaboration between an economist and a political scientist in the early stages of research into local sustainable development in different European contexts. It therefore reflects research in progress at an early conceptual stage and as such is an exploration of some of the key issues and their potential utility in future research, rather than a detailed argument for one model of empirical research and analysis. As a result the sections are organized under questions which are explored, rather than setting out an explicit method of undertaking the analysis required.

The first section addresses the issue of sustainability and argues for the need to get to grips with the dynamic nature of sustainable development and its implications. The second section discusses the terms community and local. It argues for the use of the term locality in preference to community and consideration of social as well as physical criteria in the identification of localities, in particular through analysis of relevant institutions and networks. The third section discusses the existence of several important dimensions of equity and argues for the development of an approach based on the concept of social vulnerability, which has been used to capture the multiple dimensions of inequality in other contexts of social and environmental change and which can be readily developed to explore some of the dynamic processes involved in sustainable development.

9.1 New Approaches to Sustainability

Although there is broad agreement about the need for more sustainable patterns of development, in practice it is often hard to define what this means in a specific contexts and how it could be achieved. One way forward is not to see sustainability as a fixed end state, but rather to focus on the process of attempting to move towards a more sustainable society.

The concept of a 'sustainability transition' has been developed to refer to the process of adapting to the emerging physical limits of the planet so that we can eventually live within them, whilst continuing to develop our quality of life. O'Riordan and Voisey (1997) have attempted to illustrate what some of these nascent efforts look like in the European context. The sustainability transition, as a process, is an important focus for research because different people want and desire different futures and because there is no agreement on what a state of sustainable development will be like. How different futures are defined and by whom, what the environmental and social benefits are, to what extent any consensus is achieved, who bears the costs of the changes involved and what provision for burden sharing is made are all central equity issues. There is no reason to suppose this is any less true at the local level than at the global or national level.

This can be mirrored at a policy level by emphasis by policy makers on the process, the setting in place of institutions, production processes and patterns of behaviour that will foster a less unsustainable society, rather than just the targets or end states it is hoped they will achieve.

It is necessary to be aware that the sustainability transition will happen at different speeds, in different places, to different groups of people, in a myriad of different ways. The essence of sustainable development is that it requires change in terms of our relationship with the physical resources on this planet, but this in turn requires supportive political, economic and cultural changes that address the causes of environmental degradation. Therefore we are talking about a multi-dimensional, dynamic process and it is these characteristics which need to be acknowledged specifically in any analytical approach.

9.2 What Do We Mean By Local ?

Many of the environmental sciences have a much more explicit spatial dimension to their analysis than is common in the social sciences. This can lead to the definition of local in physical terms, such as watershed, habitat or ecosystem. But where the analysis is explicitly interdisciplinary, calling for analysis of the interactions of social and physical processes, care needs to be taken to assess the relevant actors and agency in the units of analysis. It would be easy to address this simply by focusing on the relevant unit of local government such a parish, district or region. But these frequently do not match the scale of the relevant physical processes. Nor do they necessarily capture the full range of actors, stakeholders and the social processes involved. Reconciliation of the different spatial scales is not an easy process.

From some social, more people-oriented perspectives, local sustainability could be seen as referring not just to sustainability at the sub-national rather than national or international scale, but also to individual perceptions of sustainability in a specific place or social network and how the experience and choices that are made in adapting to exogenous change and moving towards sustainability, are influenced by the social processes that exist internally and externally to a locality. The processes involved may be political and cultural as well as economic.

This still requires some way of defining locality. This is by no means an easy task. Both locality and community have been used by social scientists to explain differences between places at different times (Day and Murdoch 1993). There are problems in the use of both terms, but we argue for a rejection of the concept of community in favour of locality.

Community was a heavily debated concept by sociologists until the late 1960s when it was discredited as an explanatory term. It is loaded with pre-modern, conservative ideals of idyllic rural communities, in which individuals and groups, and conflict and difference between them are subordinated to the idea of a homogenous, unique and internally consistent community that operates within society as an entity in itself (Stacey 1969; Bell and Newby 1971). However, it is still used today in much political discourse and in the fields of social policy, community and urban development. It is now also being used increasingly by environmentalists.

For many commentators, a truly sustainable society has to include action at the level of the 'community', albeit within the framework of local, national and global institutions. Thus local action and community are themes of Agenda 21 (1992) and The European Union's Fifth Environmental Action Plan: *Towards Sustainability* (1992), and in the non-governmental organisation literature (i.e. Carew-Reid et al 1994). The emphasis in these is on participation, local capacity and consensus building, local autonomy, community self-government, and empowerment as the key mechanisms for ensuring that society becomes more sustainable. Only with this kind of empowerment will local communities be able to act as effective stewards of their local environment. These aims are laudable. But whatever definition of local is used, if the analysis is to move beyond being purely descriptive or highly normative, then it must have the potential to consider what it is that has or could be enabled to have agency at the local level.

However, there is often a far from perfect understanding of what communities actually are and can do. Communities are not homogenous, but display inequalities that are important factors in the functioning and outcomes of local institutions. They very often display little in the way of consensus or the potential for it. A useful distinction has been developed between 'geographical communities', defined in terms of their physical location and 'communities of interest' which may display far greater consensus but which do not necessarily have a precise physical location.

For many practical and policy purposes the main focus of attention continues to be on geographically defined communities, because these can be matched to different levels of government and to existing targeting mechanisms at the national and EU level. If equity and effective local empowerment are to be taken seriously this poses considerable challenges for institutional innovations which will encourage the revelation of differences of interest within geographical communities and the resolution of the equity issues involved.

Could locality be a more useful concept for examining equity issues in local sustainable development? In practice there have been several approaches to locality. A locality has commonly been taken to mean a distinct place, identified in relation to other places and therefore again with a precise physical location. But locality has faced similar problems to community as a concept. Originally conceived as being less normative than community because it was spatially grounded, it has been subject to some of the same criticisms about its assumptions of social homogeneity and consensus.

However there has also been a lot of useful discussion in recent sociological and political science literature about how to characterise a locality. Some of these combine social and physical criteria in the characterization of a locality. A key paper by Margaret Stacey argued for a definition of locality as a network of institutions: so that "the locality is a context in which one can explore hypotheses about the inter-relations of institutions" (Stacey 1969:138–39, quoted in Day and Murdoch 1993). Of course this in turn requires a definition of what is meant by institution. Jordan and O'Riordan (1995) when looking at the important issue of institutional adaptation to environmental change, highlighted the many different approaches to the definition of institutions. We have followed their work in taking a wide stance on the nature of an institution. The term institution can include formal and informal institutions, ranging from social mores and cultural patterns of behaviour, to organisations and rules as set out in law.

Stacey's approach to locality can be adapted in order to look at how the sustainability transition is mediated through local institutions. She also introduced the concept of a 'local social system' to describe an 'ideal' network of institutions against which we can measure real case studies. It is possible to show that for a given place its inhabitants may engage in social relationships that in effect form a local social system, composed of the entire network of these relationships.

Such a system resides of course within a larger society and so can have many vertical links to that larger society. At the level of the individual the network of institutional links may stretch far beyond a given geographical place, so linking one place to other places. It follows that locality need not be confined to one geographical place. Indeed, for some individuals their locality may not even be grounded in one place, but being based on kinship, friendship and work links, have little reference to place at all, similar in some respects to Hannerz's (1990) cosmopolitans.

But there are likely to be some very important equity issues here, since even if it seems desirable to them, some individuals will not necessarily find it easy or possible to establish institutional links beyond a given geographical place, due for example to unemployment, poverty or the fracturing of kinship networks. In effect their social relationships may be much more confined to a single geographical place. There are a number of possible hypotheses which could be explored in more in depth empirical research.

In any case this broader approach to locality allows us to examine how social relations are embedded in place. It also moves away from favouring one type of institution over another in terms of analysis. And this approach to locality negates the assumptions of homogeneity and internal coherence which tend to accompany the concept of community, so that inequalities can be highlighted.

This is particularly useful as there are potentially large differences between social actors with regard to what sustainability means and how it can be manifested. Tracing institutional networks may shed light on these differences. In addition, this theory of locality takes into account exogenous change so that wider social, economic, political and environmental forces are not ignored. Day and Murdoch (1993:93) indicate that with this approach we can see:

> "...how the institutions in which people are enmeshed structure, and are themselves structured by, the experience of 'locality' and 'community' and also how these institutions are implicated in wider networks of relationships. It is through their participation in certain key institutions that actors are able to effect or resist change......we must be prepared to give due weight to the part played by locally integrated institutional networks in securing particular economic, social and political outcomes in given places at given times."

Some of the factors causing and affecting change at the local level at any one time will be exogenous but it could be argued that their effects are generally mediated through local institutional networks, such as family, work and voluntary groups. An institutional approach avoids both the normative or purely descriptive conceptions of locality, allowing us to look at the factors driving social and economic change, whilst taking into account particular cultural factors, such as a sense of belonging to a place or social group, local traditions, or local interpretations of change. This approach also takes into account the dynamic element of changes, acknowledging that change can occur along several dimensions, or in several institutions, at any time, and that it is often the interaction of forces for change within institutional structures, rather than just one pressure, that has to be taken into account in analyzing responses.

To sum up: observation of locally integrated institutional networks and their relationship to particular outcomes will be critical to understanding how local sustainability is defined and acted upon.

Before moving on to consider in more depth how equity issues can be examined at a local level the following points are highlighted as critical to examining local sustainability;

- Defining what is meant by 'local ' in ways that make it possible to take equity into account means moving beyond purely physical criteria to a combination of social and physical criteria.
- This will lead to a more powerful approach in analytical terms, but requires clarity of definition to move from conceptual to empirical research and for effective interdisciplinary research.
- Individuals cannot be studied in isolation from the locality, or the local social system within which they are actors, as the local context will influence their experience of change.
- An institutional network approach to the concept of locality, using a wide definition of institutions including economic, political, social and cultural, formal and informal institutions, is a way of moving beyond the definition problems we face.
- Institutional perspectives allows us to look at differing perceptions of change and vulnerability to that change, and to ascertain the resources: economic, social, cultural and political, that are available to people within the locality.
- There may be both gainers and losers from any change. The sustainabity transition is no different. The locality will be a specific context where some of the conflicts between differing needs, between the more and less vulnerable, between gainers and losers, will be most obvious and most keenly felt.
- Using an institutional network approach it is possible to conceptualize this interaction on a number of levels, as policy change and response, as behavioural and cultural change, as political bargaining and the exercise of power. In effect interdisciplinarity is allowed to occur.

9.3 Examining Equity at the Local Level

It is important to examine equity at the local level for two reasons. The first has received a lot of attention in the growing literature on sustainable development. The second has received far less attention, but could be just as important in understanding what kind of transition, if any, occurs in specific localities towards more sustainable patterns of natural resource use.

Firstly equity is an essential part of most definitions of sustainable development:

> "Humanity has the ability to make development sustainable - to ensure that it meets the needs of the present without compromising the ability of future generations to meet their own needs....Meeting essential needs requires not only a new era of economic growth for nations in which the majority are poor, but an assurance that those poor get their fair share of the resources required to sustain that growth. Such equity would be aided by political systems that secure effective citizen participation in decision making and by greater democracy in international decision making".
>
> (WCED 1987:8).

Equity is seen as mattering of and for its own sake. Greater environmental protection or economic growth which leaves equity unchanged, or even possibly worse, are not usually defined as being sustainable development.

But of course it is not difficult to anticipate a number of changes to the ways in which natural resources are used and regulated being introduced as part of more environmentally sustainable patterns of economic development, but which result in some individuals or groups becoming worse off within a particular locality. In other words groups of losers may be created during the sustainability transition. How long this effect lasts will depend upon whether they are able to adapt at low cost or to secure compensation..

Two examples will illustrate this. If subsidies to intensive farming methods were to be reduced within the European Common Agriculture Policy, this might produce general environmental benefits such as reduced biodiversity loss or pollution, but it could also lead to loss of farm income or reduced land values within certain localities.

Alternatively, the imposition of fishing quotas by the European Union could mean a loss of income or employment opportunities, reduced asset values, or more insecure livelihoods overall for fishermen and their families in specific localities where fishing is an important local livelihood. This can be traced through beyond its immediate economic impact to other effects, so that fishing quotas could also alter the social structures and cultural life of fishermen, their families and the industries that support them in a particular place and the relative political power of that economic interest group locally, nationally and within Europe.

But there is a second reason why equity could be critically important to local sustainability. Equity, or the lack of it, could have a functional significance in determining local responses to sustainability. Potential losers may not always be the poor, isolated and powerless. Indeed those who stand to lose from specific sustainability measures may be successful in preventing them from appearing on the agenda for local action at all, in altering the ways in which priorities are framed or in blocking the implementation of agreed measures in practice. Potential losers may not simply be passive victims but may be able to exercise some power through the institutions and netwrosks in which they participate. The degree and pattern of inequality existing within a locality before any sustainability transition begins may be critical to the way it does or does not emerge in that locality.

The above discussion suggests that there are three important requirements for the analysis of equity in the context of local sustainability: the approach selected needs to be able to differentiate between different groups or individuals within a locality, the analysis needs to be broadened beyond the purely economic to include political and cultural dimensions of equity as well and the approach used must lend itself to analysis of the dynamic processes involved.

This chapter proposes the following solutions: the use of the concept of livelihoods for micro level analysis of the economic dimension of equity, using the concept of social vulnerability to look at the impact of specific changes or risks and to capture some of the dynamic processes involved in sustainability and broadening the analysis beyond the individual or household to include the kinds of institutions and institutional network analysis discussed in the previous section.

Handling The Different Dimensions of Equity

The well-being of individuals and groups is likely to have several different dimensions of equity: including economic, health, political and cultural ones. Any of these might be relevant to the ways in which equity in a locality is defined externally for research and policy, depending on its specific purpose. Likewise local perceptions of equity will often draw upon more than one dimension.

But it seems very likely that the economic dimension will be critical to most definitions of equity, even if it is not the only dimension considered. It would be easy to address this through analysis of equity simply in terms of economic status indicators, such as the level of absolute poverty or relative measures, including the distribution of income or wealth within a locality. These kinds of static indicators may serve the purpose of initial descriptive analysis .But attractive though they might be in terms of ease of measurement , they will not reveal enough about how the equity issues in sustainability are experienced, defined and acted upon in specific local contexts. In particular they may not reveal enough about how people deal with the impact of negative changes on their welfare. Methods of analysis which enable the behavioural changes involved to be examined are required. Nor do simple economic status indicators always lend themselves to analysis of some of the dynamic processes involved in the sustainability transition.

An alternative starting point is the broader concept of livelihood. The term livelihood has been used extensively in the literature on sustainable development. Chambers and Conway provide a recent definition as "A livelihood comprises the capabilities, assets (stores, resources, claims, access) and activities for a means of living" (Chambers and Conway 1992). They then offered a normative definition of a 'sustainable livelihood as "a livelihood is sustainable which can cope with and recover from stress and shocks, maintain or enhance its capabilities and assistant provide sustainable livelihood opportunities for the next generation" (Chambers and Conway 1992).

Analysis in term of livelihoods offers several advantages. It can include information about the range of assets and income generating activities and not just the overall levels of wealth and income they give rise to. This makes it possible to assess the scope for flexibility in response to changes and for risk spreading. As in the definition above, assets can also be defined broadly enough to allow for both their role in markets and in other kinds of institutions to be assessed. The focus on capabilities and assets also lends itself to analysis over varying time periods, allowing some of the dynamic processes which affect the value of specific assets or capabilities to be examined.

Recent research in developing country settings illustrates the kinds of processes that can be explored using livelihood analysis. For example it has demonstrated that at the individual or household level it is not just the level of income or asset holdings that matter, but what their structure implies about the ability to spread risk. The role of income diversification as a response to variable and risky environments has been well documented in many developing country settings (e.g. Corbett 1994, 1996; Maxwell and Smith 1992). Formal economic theory has long indicated the role of holding diverse assets in a portfolio as a means of spreading risk. It follows that over- specialization in either income sources or in asset holdings may bring higher returns, but increase the risks of particular kinds of livelihood. This is likely to be very important to understanding why certain groups or individuals are able to adapt better than others to risk or changes in their physical and economic environments.

But this kind of analysis needs to be developed further through more interdisciplinary research which identifies the network of social institutions available to an individual or household within a locality. These will shed light on the role that non-economic support mechanisms and resources play in adapting to risks or changes in the physical and economic environment. This kind of analysis will make use of the full range of asset types listed by Chambers and Conway and the ways in which their value is realised in specific institutions. These might include participation in local political fora, access to education and re-training, links with other localities, familial networks as a source of childcare, social groups and spiritual institutions. Even a history of adaptation to change, resulting in a cultural understanding of change and how to cope and re-define the individual or locality within it, could be an important factor in adaptive capacity.

The resources required will not be purely economic, although the most easily measurable outcomes may be. There is much work to be done in developing the methodological tools necessary for this approach through interdisciplinary research which draws on findings from economics, political science, sociology and cultural studies.

Analysis of Social Vulnerability

The concept of social vulnerability adds to livelihood analysis by providing a tool for assessing differences between groups within a locality and their ability to respond to changes in their physical and economic environment. A summary of recent research into social vulnerability defined it as 'the exposure of groups or individuals to stresses as a result of their exposure to exogenous risks and their social and economic situation'' (Adger 1996). The concept of vulnerability has many advantages over simple poverty based analysis. This chapter argues that it is a useful starting point for the analysis of equity in the context of local sustainability, although there are several ways in which it still needs further development.

The history of vulnerability analysis illustrates some of the challenges that have to be overcome in effective interdisciplinary research. The concept of vulnerability has been extensively used within the natural sciences to examine the risks created in specific locations by natural hazards such as drought or floods. More recently it has been applied to processes of environmental and climatic change. Not surprisingly this led initially to an emphasis on physical measures of vulnerability,: such as land areas lost or risks to agricultural production. But there was increasing concern amongst social scientists, particularly in the context of vulnerability to natural hazards in developing country settings, that the focus on physical vulnerability failed to identify differences in vulnerability within the populations in affected areas. Furthermore it failed to identify the contribution of underlying social and economic factors to these differences in vulnerability.

These concerns have led to a growing literature on social vulnerability which is based on an understanding of the social and economic processes involved in making some groups within populations more vulnerable than others. For example a seminal economic contribution was Sen's entitlements approach to famines, which, amongst other things, highlighted changes in the vulnerability of different groups as market conditions changed the value of their assets and labour power, in the wake of natural hazards such as drought or flood (see Sen 1981 and for a for more recent adaptations of this entitlements approach in the context of environmental change see Leach and Mearns 1997). There is no reason why this sort of approach cannot not be extended beyond purely economic institutions such as markets. This and many other contributions have highlighted the distinction between the proximate and the underlying causes of social vulnerability.

The value of this research into social vulnerability in many different contexts is that it can encapsulate not just the initial exposure to a physical hazard or process of change, but also the extent to which social and economic factors intensify the risks involved for particular individuals or groups within populations.

Recent research has even applied the concept of social vulnerability has been applied to other contexts in which the immediate sources of risk and stress are economic rather than physical, such as the rapid macro-economic change resulting from the adoption of structural adjustment programmes in indebted countries . This is relevant since the sustainability transition is likely to involve a number of economic policy changes to reduce risks to the physical environment.

In economic terms social vulnerability has increasingly been defined as disruption to livelihoods and/or loss of economic security at the level of the individual or household. But this does not go far enough in taking into account the factors which influence social vulnerability, including political power, education and skills and gender roles. Again research in many developing country settings (summarised for example by Chambers 1989) has shown that when open ended questions are used to solicit the perceptions of those involved vulnerability is experienced subjectively as being political and cultural as well as economic. There is no reason to suppose that this will not equally be the case in European settings.

One of the advantages of the concept of vulnerability is that it includes the idea of risk and the need to respond to changing and often uncertain conditions. Rather than a simple static analysis of who is adversely affected by a given change initially, this leads onto the analysis of the ability of different groups or individuals to adjust to adverse changes and the processes involved. Why are certain groups better able to adjust and eventually to adapt than others ? This is particularly relevant to the dynamic concept of a sustainability transition. It can capture uncertainty about the scale and pace of different kinds of environmental change and the risks involved, and makes it possible to take a more behavioural and institutional approach.

Again, research in developing country settings has been useful in developing the concepts of concepts of coping, adjustment and adaptation to risks and stresses (Corbett 1988; Mortimer 1989 and reviewed in Maxwell and Smith 1982) to examine how vulnerable individuals respond to adverse changes and/or try to reduce their initial vulnerability. A useful illustration of this is a common finding in the literature on vulnerability to drought that even in very poor populations, individuals who are exposed to repeated risks, are likely to develop some kind of insurance behaviour to reduce the risks involved and that informal means of self or collective insurance may still be developed in the absence of formal insurance markets (e.g. Corbett 1988). In the context of novel or extreme events the critical question is how quickly these kinds of behaviour or institutions emerge, given that insurance of any sort usually has an opportunity cost in terms of the resources committed to it. Indeed some might argue that the emergence of sustainable development as an organizing principle for action can be seen as a form of insurance for mankind as a whole.

The critical equity questions which social vulnerability analysis can shed light on are : what determines differences in the ability of different individuals to respond to the stresses and disruptions to their livelihoods which are imposed by exogenous changes? Why do some people and local social systems fail to cope, whilst others are able to adapt successfully without long term loss to their economic or social status? Of course in analysis of adjustment to sustainability measures in a locality it is clearly important to determine how far the factors that are critical to coping or adapting are themselves determined at the local level or alternatively dependent upon resources or opportunities from outside the locality.

The analysis of livelihoods and vulnerability needs sensitive interpretation . There is no reason to suppose that all individuals or groups will be able to enjoy secure and unchanging livelihoods. The concept of sustainable development does not necessarily imply simple preservation of fixed economic or social states. Even without environmental change or agreement on the need for sustainable development, many other factors may require livelihoods to be in a continuous process of change and adjustment, requiring skillful management by the individuals or groups concerned. Likewise, equity does not need to mean the preservation of a fixed structure of livelihoods in a specific locality, but it does need to take into account the costs: economic, social and cultural, involved in adjusting and adapting livelihoods.

Social Vulnerability and the Dynamic Nature of Sustainability

As indicated earlier it will be very useful to use a dynamic approach when analyzing sustainability, because it is as much a process as a state. In this research it is assumed that there may be significant differences in the paths to greater sustainability adopted in different localities and in the speed in which change occurs.

At the local level it may be possible to make progress on some issues on which there is a fair degree of consensus of the need for change, particularly if they are considered amenable to change beginning at the local level and the costs involved (either economic or social) are seen to be relatively small. Change in other areas may be much slower to emerge or contingent on other processes.

The vital point that will emerge when vulnerability analysis is conducted is that it will not always be the same groups or individuals who are vulnerable at all stages of the sustainability transition. Initially three different stages might be envisaged: before the transition during and after. This of course is an over simplification, but it will serve to illustrate possible changes in the identity of vulnerable groups at different stages of the move towards greater sustainability. Several different profiles of vulnerability can be envisaged.

In the first case, those who are very dependent on intensive use of natural resources as the dominant source of income in their livelihood, such as the fishermen discussed above, may not see themselves as economically vulnerable initially. They may be unaware of any adverse environmental change or irreversible resource depletion resulting from this kind of livelihood. Even if they are aware, they may still see it as economically rational at an individual level to continue to fish at unsustainable levels.

But, they may start to consider themselves as vulnerable as either local or more general measures are developed to encourage different and more sustainable patterns of resource use, if these could result in rapid loss of income or other disruption to their livelihood, without alternatives being readily available. In this case as a result of their perception of their increased economic vulnerability they may wish to block the implementation of these measures, or be compensated, in ways which could be very important to the scale or pace of the sustainability transition in their locality.

But, there are other possible vulnerability profiles during the sustainability transition. Some people, and again those dependent on fishing livelihoods are used as an example, may already be suffering from declining incomes due to levels of over-fishing which they are individually powerless to reverse. They may already perceive themselves to be economically vulnerable before any new measures are introduced. The introduction of collective measures to manage fish populations more sustainably may bring both greater security and sustainability to their livelihoods and therefore a perception of reduced vulnerability. However these measures may also reduce the total number of such livelihoods that are possible, causing a large short-term adjustment which may be spatially concentrated and experienced particularly severely in some localities due to the homogeneity of livelihoods within them.

And thirdly there may also be other groups who were already vulnerable before the transition begins, for example due to asset poverty, low incomes or insecure livelihoods overall and who remain vulnerable even as measures begin to be implemented or whose vulnerability is exacerbated still further by them. In this context the concept of 'baseline vulnerability' used by Watts and Bohle (1993) is useful.

Of course it will not always be easy to separate out the different determinants of social vulnerability and the interactions between them: thus presenting a number of challenges for research in specific localities. In the European context, for example, the impact of economic integration through the creation of a single market may be altering patterns of employment and the structure of livelihoods in different places. European wide policy to address specific kinds of environmental change or to implement sustainable development measures may be happening at the same time and without co-ordination with local action to promote sustainable development.

For example, in the dairy farming area of South Devon the combined effects of the milk quota system in the Common Agricultural Policy and the landscape and biodiversity conservation measures imposed by European Directives and Government designations have meant farming practices have altered the local economy, farmers and support industries have been forced to diversify, particularly into tourism, thus altering the social profile and agricultural culture of the area (Voisey 1996).

To summarise: given that current thinking about sustainability assumes that there can be and needs to be significant agency and action at the local level, it is vital to be aware that the identity of losers may be different at different stages of the sustainability transition. Furthermore, where potential losers are not the poor, isolated or powerless but rather wealthy or politically powerful groups, they may be able to affect policy priorities and political agendas and what kinds of compensation or assistance with adjustment are provided to allow change to occur.

Equity and Extending Social Vulnerability Analysis

The term vulnerability implies a focus on those who are adversely affected or lose from a specific change. But it is possible to envisage many changes which result in both gainers and losers. For example within a locality a change from a polluting or energy intensive industry to more environmentally friendly industry may result in both losses and gains in employment, with individuals with redundant skills and experience losing whilst others gain. The term differential vulnerability needs to be used in ways that can include both gainers and losers if the full dynamics of sustainability within a locality are to be examined.

It is also essential to be clear about where the losers and gainers are in relation to a specific locality. In some cases both the losers and gainers may be found in the same locality, leading to a redistribution of welfare within the locality .

But it is also possible to envisage other situations in which most of the gainers are outside the locality and most of the losers within. An example might be measures for increased protection of a particular habitat or ecosystem with national, regional or global conservation value. This kind of change has implications for the relative power of local interests to those externally in terms of the extent of compensation which is brokered.

Finally and given the multidimensional nature of human welfare and therefore of equity, it is quite possible for the gainers and losers to be the same individuals, groups or localities. Again, using the example of the closing down of an environmentally polluting industry , in the short term this will adversely affect some livelihoods through loss of employment, but it may also improve health conditions within the locality. This illustrates that it will often be necessary to differentiate between different dimensions of vulnerability and not to use single dimensional indicators alone.

Limitations of Locality Level Analysis

Finally an underlying issue that needs to be reviewed is whether all the equity issues in local sustainable development can be addressed through analysis at the local level alone. The answer is clearly no. Although it is very important to assess what is happening within a locality, it is unlikely that this will be sufficient. Many of the measures undertaken to promote more sustainable patterns of development may be determined at the national, regional or even the global level. There will also often be other social and economic factors which affect baseline vulnerability within localities, but are not determined at the local level, such as the impact of national or European Union level economic development policies. Even where there are opportunities to access or redistribute resources and participate in decision making at the local level, it is important to examine the national, European or global context within which these occur and to differentiate between exogenous change and change initiated within a locality. Analysis of institutions and institutional networks provides a way of examining the interactions between the two.

The issue of burden sharing in the sustainability transition is also very relevant. The term burden sharing is often applied to international contexts. But it could equally be applied to analysis of losers and gainers at the local level. But it will not always be easy to make simple assumptions about patterns of burden sharing at the local level during the sustainability transition. Several possible patterns could be envisaged as indicated previously. They could be both losers and gainers locally, but also groups external to the locality who can impact on those internally. In many cases it will simply not be possible to spread risk adequately within the locality where losers are concentrated within it. Measures to compensate losers may often involve the use of resources external to the locality through interaction with national or European institutions.

9.4 Summary

This examination of the potential significance of equity issues at a local level in the context of sustainable development has identified the following research issues as important:

- Locality has more utility than community as an approach for analysing social interactions in a specific place, particularly the institutional network approach devised by Stacey, which provides an analytical tool for looking at inequity, forces of change, methods of adaptation and social support mechanisms within a locality.
- Equity has a functional significance as well as mattering for reasons of social justice. The extent of and pattern of inequality within a locality may shape the speed with which a sustainability transition takes place, as well as the path it takes.

- Equity has several dimensions, including economic, political, cultural and health ones. Although far from a complete description, the economic differences between groups within a locality are likely to be critical to most analysis of equity.
- Analysis of the structure and distribution of livelihoods within a locality provides a useful method for examining micro level economic equity issues. However this will often need to be broadened by an assessment of the role of non economic institutions in determining the structure and security of the livelihoods of different groups within a locality.
- Examination of how different localities attempt to make the transition to greater sustainability and tackle the equity issues involved requires a dynamic approach to equity. Because of its emphasis on risk, change and stress, the concept of social vulnerability offers more scope to do this than simple indicators of absolute or relative poverty.
- Recent research on social vulnerability in other contexts suggests that the behavioural concepts of coping and adaptation are important in examining how individuals or groups respond to exogenous changes and the costs of doing so.
- It should be possible, using analysis of locality and social vulnerability, to highlight what adaptive processes are being followed, where new processes might be developed, or policy responses required to improve adaptive capacity, either for certain individuals, groups or localities.
- Vulnerability is not the same as poverty. Those who are most vulnerable: due for example, to dependence on risky resources may be vulnerable because of the specialised nature of their livelihoods. But they may also be relatively wealthy, affecting the nature of compensation society provides. Where potential losers are also powerful then they may be able to block the implementation of sustainability measures or even alter the ways in which it is defined.
- Local perceptions of who is socially vulnerable and why, may be based on different kinds of information and criteria than the theoretical ones identified by outside researchers. In interpreting responses to specific changes it will be important to investigate these local perceptions and to acknowledge their potential sensitivity to respondents.
- The term social vulnerability implies that the only equity concerns are potential losers from sustainability measures. But there may be both gainers

and losers from specific measures to protect the environment or reverse environmental change processes. Again an interdisciplinary approach will be useful to an analysis of the interactions between gainers and gainers.

Although, it will be useful to do detailed micro level analysis at the local level, it is also vital to situate this within the broader economic and political context . Many of the determinants of baseline vulnerability may be due to factors outside the locality. Furthermore the scope for successful adaptation may depend as much on the vertical linkages that different individuals and groups have beyond the locality, as on the opportunities and networks within it. These vertical linkages may be critical to risk spreading , through access to resources, new markets, institutional networks and policy coalitions outside the locality as well as within it. More detailed analysis of the response of different localities to the challenges of local sustainable development is vital, but should not be over deterministic.

Conclusions

This chapter has been concerned with how the process of moving towards a more sustainable society can be observed at a local level. This has involved consideration of the most useful definition of local and the ways in which equity or the lack of it are experienced at a local level in the context of sustainable development. The challenge has been to develop an approach that is both informative and tractable as a basis for empirical research.

Equity is a central concept to the idea of sustainable development, but making it operational is problematic. This is in part due to failure to see that inequity is not just a comparative notion looking at one variable, rather it is multidimensional. It is this aspect which is highlighted by a holistic focus on the locality as the site of analysis. If equity is to be achieved as part of sustainable development, then it is important to recognize that people experience change differentially, both between and within geographical places and that this mediated by economic, social, political and institutional variables at a local level. A concept of locality, as well as social vulnerability, which brings this to the fore will help to illuminate what is happening or should happen when looking at progress towards sustainability.

The approach proposed involves analyzing equity issues at a local level using the concept of social vulnerability in a more multidimensional and interdisciplinary way than is currently used. This involves analysis of livelihoods, local institutions and institutional networks and their linkages with the broader economic and political context. The enormous cultural, ethnic, social, political, economic and institutional diversity within a region like Europe, requires a framework for analyzing the interactions of these spheres that moves away from description and comparison towards a much needed interdisciplinary effort to get to grips with the dynamic nature of sustainable development and its implications. Here we have attempted to illustrate how such an approach could be constructed, although much conceptual and methodological work still needs to be done to develop the framework presented.

When and if it happens, the sustainability transition will happen at different speeds, in different places, to different groups of people, in a myriad of different ways. The essence of sustainable development is that it requires a change in our relationship with the physical resources of the planet. This will require supportive political, economic and cultural changes that address the causes of environmental degradation. This implies a multi-dimensional and dynamic approach and it these characteristics which need to be acknowledged in the development of any analytical approach.

REFERENCES

Adger N. *Approaches to Vulnerability to Climate Change*. Centre for Social and Economic Research on the Global Environment (CSERGE) working paper 1996, GEC 96-05.

Bell C, Newby H. *Community Studies*. London: George Allen and Unwin Ltd, 1971.

Blaikie P, Cannon T, Davis I, Wisner B. *At Risk, Natural Hazards, People's Vulnerability and Disasters*. London: Routledge, 1994.

Carew-Reid J, Prescott-Allen R, Bass S, Dalal-Clayton B. *Strategies for National Sustainable Development: A Handbook for their Planning and Implementation.*Earthscan in association with IUCN and IIED, London, 1994.

Chambers R. Editorial Introduction: vulnerability, coping and policy. IDS Bulletin 1989, 20:2,1-7

Chambers R, Conway G. Sustainable rural livelihoods: practical concepts for the 21st century. IDS Discussion Paper 1992:296

Commission of the European Communities. *Towards Sustainability: A European Community Programme of Policy and Action in relation to the Environment and Sustainable Development*, COM (2) 23, 27.3.1992.

Corbett J. Famine and Household Coping Strategies. World Development 1988; 16:9:1099-1112.

Corbett J. *Livelihoods, Food Security and Nutrition in a Drought Prone Area of Zimbabwe*. Final Research Report to ODA/ESCOR. January 1994.

Corbett J. *Dynamic Processes in Food and Livelihood Systems: Links with Ecological Resilience.* Final Research Report to ESRC/Global Environmental Change Programme, 1996.

Day G, Murdoch J. Locality and community, coming to terms with place. The Sociological Review 1993; 82-111.

Dow K, Downing TE. Vulnerability Research: Where Things Stand. Human Dimensions Quarterly, 1995;1:3:3-5.

Hannerz, U. Cosmopolitans and Locals in World Culture. Theory, Culture and Society 1990: 7:2-3 237-253.

Jordan A, O'Riordan T. *Institutional Adaptation to Global Environmental Change (I): Core Elements of an 'Institutional' Theory.* CSERGE Working Paper, 1995; GEC 95-21.

Leach,M, Mearns R. Environmental entitlements: a conceptual framework for understanding the institutional dynamics of environmental change. IDS Discussion Paper 1997: 359

Maxwell S, Smith M. Household food security: A conceptual review in *Household food security: concepts, indicators, measurements: a technical review.* Maxwell,S.and Frankenberger. UNICEF/IFAD 1992.

Mortimore M. *Adapting to Drought, Farmers, Famine and Desertification in West Africa.* Cambridge: Cambridge University Press, 1989.

O'Riordan T, Voisey H. (eds.) Sustainable Development in Western Europe: Coming to Terms with Agenda 21. Special Issue Environmental Politics, 1997; 6:1.

Parker J, Selman P. *Policy, Process and Product in Local Agenda 21, Working Paper no. 2: Interview synopses: Gloucestershire.* Cheltenham: Cheltenham and Gloucester College of Higher Education, 1996a

Parker J, Selman P. *Policy, Process and Product in Local Agenda 21, Working Paper no. 3: Interview synopses: Lancashire.* Cheltenham: Cheltenham and Gloucester College of Higher Education, 1996b.

Sen A.K. *Poverty and Famines.* Oxford: Clarendon, 1981.

Stacey M. The myth of community studies. British Journal of Sociology, 1969; 20:134 - 147.

UNCED. *Agenda 21.* http://unep.unep.no/unep/partners/un/unced/agenda21.htm, 1992.

Watts M. On the poverty of natural theory: natural hazards research in context. In *Interpretations of calamity* Hewitt K (ed) Allen and Unwin: Boston: 231-262.

10 CULTURAL EMPOWERMENT :

(Re)building Locality and Facilitating Collective Vision as Interventions Toward Sustainable Rural Development

Nicholas Mack

Introduction

'Place equals space with meaning' (Clifford and King 1989)

Might culture, locality and rurality provide a basis for an intervention strategy for sustainability?

It is not unreasonable to suggest that alongside an individuals' economic concerns, his or her actions are defined by social and cultural influences as well (see for example, Bourdieu and his concept of 'Habitas' 1977). What is difficult, however, is finding a means to get an operational handle on how social and cultural influences come to operate on individual behaviour, (there are so many different sources for them). Even more difficult is envisaging how cultural influences might be intercepted and harnessed within a specific strategy of intervention intended to furnish situations of sustainable development.

As they stand, ecological / economic system models which are designed to facilitate the goal of sustainable development, and which attempt to build causal bridges between ecology and economy, between society and nature, are based largely on observable and hence measurable phenomena – essentially, peoples (consumption) activities and the environmental consequences of these. Culture is invisible – assumed into the background. But why should this be a problem? Does a lack of 'culture' in the equation make ecosystem models inadequate?

Perhaps it does; because in this way human behaviour tends to become 'simply' a matter of economic rationality. But even if these models were sufficient to predict economic/ecological relations without involving 'culture', leaving it out may ultimately make them impotent. Without being able to 'empathise' and engage with, indeed, recruit, people-as-actors, recommendations on policy and practice may be limited. It may not be the models which are our concern, then, but the implementation of their recommendations.

As such, I do not intend to suggest how 'culture' might be measured, which would be a foolish task. Rather, I wish to introduce, assess and explore the role of a cultural dimension in the practical realisation of sustainable rural development. To do so I will go beyond the traditional notion of culture in rural development; one of community festivals and 'local crafts', and beyond the idea of 'culture-as-PR' , of trying to persuade people to change their ways through media image.

My suggestions for the use of culture will instead begin by presenting apparently simple things in a somewhat complex way. Beginning with an abstracted view of culture is intended to focus attention on its role in forming meaning and informing experience, as a multi-faceted and synthesising process of communication and information exchange. It anticipates the need to render sustainability (and its various indicators) visible in day to day life - to seek to 'develop'; more in the literal, photographic sense, a consciousness within society of 'being sustainable' and in a way which is in keeping with the complex system dynamics of whole ecologies and whole societies, woven together.

This presentation of culture is therefore cybernetic (Conway 1987), that is, capable of seeking an equilibrium optimal state which holds together complex system information. Culture becomes the carrier of complex information, providing a holistic, that is; intuitive as well as rational, means to build sustainability within social activity and decision-making. It is a means which goes beyond the positivist and cumbersome academic/professional capacity to monitor and manage sustainability alone.

From this abstract working up of a cybernetic model of culture I will suggest where a strategy for intervention can be rooted. I will then review some of the realities and challenges confronting such a strategy. Having done so, I will report on some current and upcoming practical initiatives which begin to address these challenges, and then return to review the prospects and parameters for formulating a cultural strategy for sustainable development.

10.1 Some Conceptual and Theoretical Parameters

I will begin, then, by providing some broader conceptual and theoretical parameters for 'culture', for 'believing in' a cultural strategy in sustainable rural development. There is, I must report, an immediate problem to be overcome. Even amongst the anthropologists, to whom we might have turned for a more definitive answer, what culture is or means is far from having been agreed, (Milton 1996). Drawing from the recent work of Anthropologists such as Milton, though, along with Social Psychologists such as Shotter (1995) and a broad range of work elaborating cultural factors within Rural Sociology, a framework for apprehending 'culture' in society can be developed. It is one which focuses on the continuous generation of meaning in society; the constant interrogation and interpretation of reality by all individuals together and independently as a guide to action. To begin, we might unpack 'culture' into two broad components, loosely termed 'product' and 'process'.

Product is 'historic' – made up of accumulated and recorded knowledge and experience. Some of this may be invoked by and will inform contemporary 'process' (below), whilst some may have become unused and largely forgotten. Product is inherent in social norms, social and cultural capital, and patterns of legitimacy – in these things it pre-figures the distribution of social power.

Process is 'creative' – culture is a component of communication and the (re)creation of meaning – implicit in everyday conversation and public discourse and explicit in public events, arts, crafts and festivals, and so on. Some of what appears to be 'process' is in fact 'product' – a reification of established forms and traditions. Process requires a degree of re-invention – or of being 'lived'. It can challenge and overturn established 'product' replacing it with new.

Process, then, incorporates ongoing experience and describes / interprets / forms representations from it, to present to others. Presentation may be intended to persuade (Callon 1986) or to maintain (Cohen 1985) or to generate knowledge (Shotter 1995). Process re-works 'Product' but is also governed or regulated to variable degrees by it – in effect, process and product interact to generate a cultural 'evolution'; the basis of a cybernetic model.

The general activity of interpretation, representation and communication in whatever form, verbal or artistic, formal or informal, is of particular interest and will be collectively described as discourse – discourse is

> 'not just another word for conversation ...it refers to all the ways in which we communicate with one another, to that vast network of signs, symbols and practices through which we make our *world(s) meaningful to ourselves and others*'.
>
> (Gregory 1984 p11)

As Milton puts it 'Culture situates us in the world...and it is through culture that we reflect on our actions and experiences, describe them to others, and plan future courses of action'. I will wish to introduce the idea of a 'rural aesthetic' as a particular set of signs and symbols available to social discourse which is a basis for a cultural sense of sustainability.

Discourse is not uniform but takes place in different 'fora' or domains. Adapting the broad fields of discourse identified by Jones (1995), three types of domain are of particular interest here, which we might term :

- Academic/Professional ('The Top')
- Wider Societal or Popular (The 'Middle')
- Local ('The Bottom')

In principle, each of these could be broken down into other 'sub domains' - those of the individual academic disciplines, those which are formed by policy makers, or by development agencies, and so on, which together make up the 'Top' . In 'the Middle' we might discern domains characterised as youth media, business media, popular media, and so on. These distinctions are not helpful though for present purposes, in looking at what broadly influences the shape and nature of 'sustainable rural development'. It is only helpful to create such distinctions when we come to look at the way 'The Bottom' is constituted, in terms of its various 'communities of interest' within any given local geographic 'community'. Farmers, business interests, younger people, mothers, grandparents, will all have particular relationships with and experiences of the local-as-place which reflect different needs, values and personal histories.

Each of the three broad domains I have described are elaborated by a particular sort and structure of discourse - the boundaries distinguishing ones position if passing through these three different domains would be defined by differences required in the use of language, in the rules of discourse and of other specific symbols. Each has developed a specific context or accumulated product – a different knowledge base in which and from which interpretation and representation takes place. Each 'domain' therefore sustains, in effect, its own culture, and the invader from one to the other becomes easily recognised.

The 'products' of discourse within each domain, however, go on to influence subsequent discourse in each of the others, forming; 'flows of meaning and power which combine to create social constructions of such things as the rural' (Jones, ibid). The interactions are complex, but again, broad 'channels' or interfaces can be suggested where 'translation' takes place from one domain to another. Using rural development as the example, a set of interfaces can be defined which together form what I shall term a 'discourse triangle'.

For example, a key, obvious, and well studied interface is that between academics and professionals on the one hand (the 'Top') and 'local' people or actors on the other (the 'Bottom'). This particular kind of interchange is about influencing the direction of rural economic and social developments and technologies. The translation is obvious, from 'abstract' to concrete, for example. It can be labelled 'Top-Down', in describing the predominant type and direction of flow of meaning, meaning which is 'strategic' in nature and has its home in 'the Top', flowing into local domains through policy instruments and extension work. Even whilst present day rural and community development work has a 'Bottom-up ethos', in practice the 'Bottom' is dominated by the culture of the 'Top'.

A second key interface is between local communities and wider society. This interface is more 'diffuse' and unstructured, conducted variously through the linkages and networks different people have. In pursuing their various activities, individual contacts stretch from the local area they each share into different 'scales' of society and geography (Cox and Mair 1991). Much is mediated through the auspices of the marketplace and of the media. The trend for translation between domains, though, is from the larger scale one of societal discourse into the practices of local life. It may in part inform development of a 'service' local rural areas could be said to have of providing for 'rurality' in society at large - in the extent to which rural dwellers act as stewards, or in practising animal welfare, or in the design of tourism experiences and so on. But it is also and more predominantly, spurring individuals toward reducing the disparities they perceive between their own, localised circumstances, and the style of life portrayed for society at large, moulding or marginalising 'the local' in the process. Again, 'the local' is dominated by the culture of the 'Middle'.

A third interface is between wider society and the 'Top' - but to the extent that public opinion might be said to shift policy and development practice, the result is largely realised in relations between the 'Top' and the 'Bottom' - in policy and extension work. Examples include the extent of pesticide use recommended in farming, or the encouragement of conservation practices. Likewise, where science, technology and policy themselves affect society at large, through new practices, there follow ramifications for the impact of society on rural areas, and the uses society wishes to make of rural areas.

The result can be depicted in the form of a 'discourse triangle' :

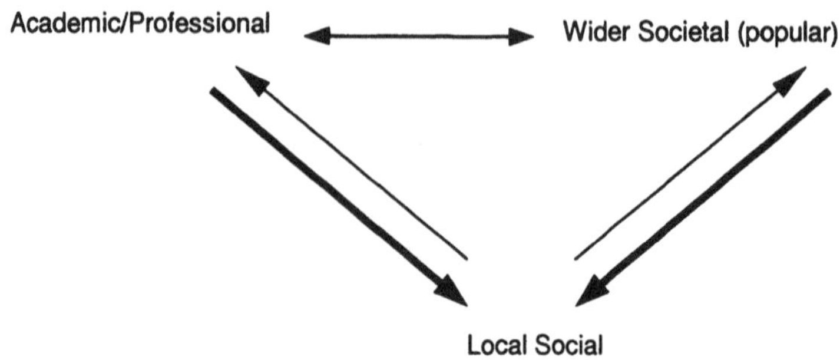

In this triangle, it is the local which represents a key meeting point in the flows from the 'Top' and the Middle', particularly when the subject for these discourse flows is 'rurality'; that is, the interaction between society and ecology. Yet the weakest point in this triangle is the 'local'. There is little scope or capacity for 'the local' to feed back a recognised discourse to society or to the 'Top'. Indeed, the very idea of 'the local' is a weak concept when subjected to scrutiny - its definition is blurred and uncertain, as we shall see.

10.1.1 The Cultural/Ecological Interface of Local

I mentioned that Discourse works within 'contexts' or domains each made up of an internal knowledge base or elaborated internal 'culture' distinguishing one from another. The 'local' is perhaps unique in this regard in having a physical dimension informing its context, and in the rural local, this physical context includes natural / ecological systems and entities.

The 'local domain' is the one which will be most informed by this physical context - not, by any means exclusive to other social influences, but certainly in a way which is more pervasive and continuous, and less abstracted, than will be the case in the other two 'domains'. It is also the place at which the human 'cultural discourse' becomes translated into physical constructs, and interacts with ecological systems. The ecological or bio-geographic is the very 'Bottom', of 'Bottom-up' – found, as it were, underneath the discourse triangle[1]. It forms the stuff of which rural is actually rendered – landscape, forest, and rivers, lakes, mountains, and wildlife and the spaces and perspectives which relate them. These have the least capacity of all to reflect back into the triangle. The full version of the triangle which includes this bio-geographic interface then looks like this :

[1] 'Underneath' in terms of being 'subject to'

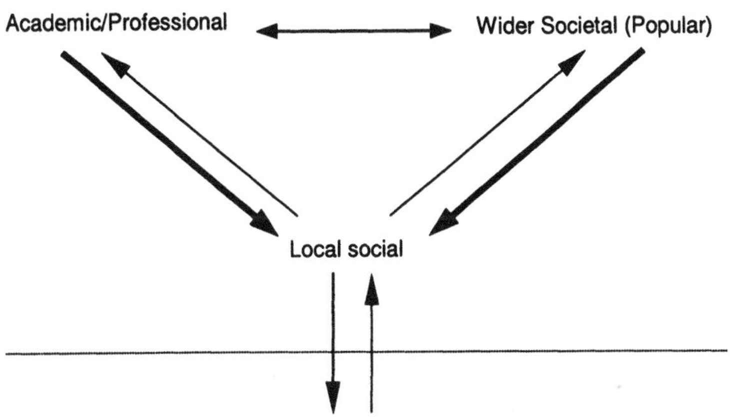

In this full version of the triangle, there are pathways evident between not only different cultural domains, but also the physical and bio-geographic domains as well, which intersect at 'the local'. The local becomes a point of arrival.

10.1.2 Beginning To Define an Intervention Strategy

Those with a background in rural sociology may at this point begin to fear the resurrection of an old and discredited romantic idea; that of the rural local community, in touch with itself and its surroundings, captured by the term 'Gemeinschaft' as far back as 1887 by Toennies, and only recently exorcised by rural sociologists during the 1960s and 70s. I will address the difficulties of appealing to concepts such as 'local' and 'community' during subsequent parts of this chapter. Introducing a strategy of 'cultural empowerment' obliges some aspiration to engender 'gemeinschaft' or sense of place as a social resource, but this does not also imply that 'Gemeinschaft'; 'community-in-place', is a distinctly rural characteristic which can be set apart from a contrasting detachment and individualism seen for urban society.

I do, however, wish to distinguish the rural local by its substantial element of natural, bio-geographic and ecological features. This is incidental to the basic principle of a physical setting providing unique contexts to a local discourse - a built environment will also do this, but the bio-geographic nature of 'rural' becomes very important for Sustainable Rural Development to recognise and address. An intervention strategy designed to open up a cultural pathway to sustainability should, I will suggest, seek to amplify this bio-geographic aspect of the rural physical setting for society at large, through building a local collective vision or consciousness of natural resources. What is different in these proposals to the original idea of Gemeinschaft is that we should not pre-suppose that this consciousness exists already among rural populations.

A key to establishing sustainable rural development then, may be to 'empower the local', not in the ways usually associated with this term in rural development, in economic terms, or in terms of local communities influencing the planning processes (important though 'participatory planning' will be to sustainable rural development). Rather, I would propose empowerment in 'cultural' discursive terms. Such a cultural empowerment perspective is not restricted to enhancing only local social discourse, but also, in so doing, seeks to more fully enable an influence or presence from the local bio-geographic base within and through this discourse. In effect, a cultural empowerment perspective, seeks to empower the ecological, through the cultural!

This empowerment should take place by actively working on strengthening the contextual knowledge base, that which provides the distinctiveness and integrity of each of the three main discourse domains identified in the triangle. In the case of the local domain this means recognising and amplifying the physical element. By so doing, the local domain, and the ecological domain, are enabled to generate outputs influencing the discourse triangle, for example, in the production of a rural aesthetic[2]. The aim is to foster an enhanced, cybernetic approach to sustainability.

This chapter and the practical initiatives it describes, then, is aimed at exploring and implementing a 'cultural empowerment perspective' as a means of acting to capture, translate, and reflect discourse about the environment and rurality, in a manner which does not subjugate but rather, establishes an equality of and validity to the local discourse domain shaping rural development and rurality in society. It does so incorporating ecological and bio-geographic meaning, and in a way aimed at establishing a level of collective vision or sense of place across different local interests.

Work current within Scottish Natural Heritage and the Rural Development Council is seeking to develop the tools, techniques and frameworks in which to use a cultural empowerment approach, and to build partnerships toward sustainable rural resource management, but before looking at these I want to address some of the problems which might be anticipated for such a strategy.

---

[2] This rural aesthetic may be one way to capture and display the complex information about interactions between social, economic and ecological systems and to demonstrate the condition of that interaction – how well might it imply sustainability in holistic or intuitive terms – in cybernetic terms ?

10.1.3 Problems with the Theory

In invoking the terms 'local culture', and rurality, and seeking to build a cultural empowerment process, we are faced with a number of problems.

Firstly, to what extent or in what conditions can we expect or hope to find a 'sense of being local' in a globalising society? This is a critical resource from which to build a collective vision toward cooperative action. There is an ongoing debate about the importance (or not) of 'Place' and what meaning it has, a debate which has been occupying the thoughts of sociologists and geographers for some time (see for example Massey 1994, or Cooke 1990).

Even where there is a local sense of 'place', how vibrant is local 'culture'? Can it be disentangled from cultural values held and articulated in wider society? (does it need to be?) In effect, can the 'local' point identified on the discourse triangle be distinguished or realised as a discourse domain, and built upon? A globalising society may not only encourage detachment from place, but a reduction in the discourse the local attracts – the attention and value it is given as local, eroding the means for holding 'the local' together as a recognisable experience through the increasing trend to individualisation.

Secondly, what does the local look like in terms of size and social relations, and how might it operate in achieving sustainability and collective action? How big is 'the local' – does the social fit with the ecological?

Thirdly, the term 'rurality' itself has come under attack from a number of sources. The discussion in this chapter is particularly concerned with 'Rural Places'; those in which there are multiple points of contact between the ecological, the social and the economic, which can be crystallised by the cultural. This is perhaps the least of our problems because increasingly, 'Rurality' is being viewed as a cultural ' discursive' product.

I will look at each of these 'problems' in turn as theoretical issues, but judgement about the real situation of each must largely be reserved for the outcome of the practical action research initiatives and the development of suitable development tools outlined in the last part of this chapter.

10.2 Practical Difficulty Number 1: What is rurality anyway? Does it have any currency?

Our relationship with nature has been said to be irreversibly changed by the emergence of industrial capital, so that

> "the reality of 'social domination' over nature is given, even if the extent of the process is a matter of debate and its morality the object of bitter struggle"

(Smith 1990, p1)

What management value or ascribed local value can 'cultural' interpretations and relations with nature have in such a condition of domination?

Smith suggests that a dualism has informed the development of modern society, one between 'external nature' and internal human nature, which can be traced back most directly to Kant. Positivist science as propagator of industrialisation has tended to treat 'nature' as external. The industrial, neo-classical labour process treats natural materials as external objects of labour to be transformed into economic commodities. This 'externality' is implicated in Noorgards' (1987) critique of Neo-Classical Economics and its ability to accommodate natural systems.

Yet there remains a stubborn 'universality' of nature which includes us as part of it, which we also , in our hearts, recognise. *"Nature is somehow both external to man (sic) that which is not man, and it is man as well as nature"*. (Smith, p7) 'Man' is also a manifestation of nature, then, and whilst the 'positivist project' has done much to create boundaries between 'man' and nature, it is yet difficult to deny the experience of feeling part of it as an informing principle of the human condition.

If it should be doubted that such a feeling exists amongst us; of ultimately, being part of nature, we can turn to the increasing visibility of environmental activists, their determination, and the sympathy they now have in the public eye. These activists are emerging over a wide range of issues and include people from a broad range of social classes, backgrounds and interests. The 'battles' being fought, in tunnels and in the trees, at ferry ports and in the North sea, show how important the 'well-being' and availability of nature is felt to be.

Alternatively, among 'objective' academics, we might look at the substantial difficulties faced by rural sociologists (amongst others) to satisfactorily explain or deny the essence, the nature of 'rurality' as a causal factor. Whilst its reality is objectively elusive, nevertheless it has not been possible for the idea of the rural to be 'Got Rid of' (Hoggart 1990). In a similar vein, the growth in efforts to elaborate and operationalise the concept of sustainability provides further clues that the domination of 'nature' when viewed as a complex system is not so complete, nor its effect on our psyche a thing of the past.

For Kant, the dualism between 'inner beast' and outer nature was to be overcome by culture. His 'inner beast' is undoubtedly a reference to the need for culture to tame the our baser instincts inherited from nature, but perhaps we can also equate such instincts with the 'pure economic act', the competition for resources. Culture, however, which here becomes a space in which to insert a moral mediation to such acts, is a quality which eludes positivism. As an intermediary space between motivation and action, it eludes causality, and as such, it rarely features in our efforts to model or understand individual behaviour.

Again, the work of rural sociologists illustrates this elusiveness. Their bid to understand 'rurality' and its causality has taken the 'linguistic turn', to analyse it as a cultural/experiential process in the collective conscious of society. (see for example Halfacree 1993, or Mormont 1990). It has turned to social psychology for what have been termed the 'social representations' of the rural; in language, in art, in the media. Elaboration of this sort (focusing more often than not on the expression of everyday culture through conversation rather than the 'grand expression' of artists – but see Matless 1995 for a combination of sources) is aimed at translating 'cultural expression' into a scientific format which can be analysed and hence given more robust form; and perhaps, validation.

Social representation theory as portrayed by Moscovici (1984) suggests a common 'pool' of ideas and values, of mental images, which are drawn upon, and then translated in or applied to local contexts. The pool of values about the rural are at one level 'in potential', therefore, and exist across society as a whole in latent form (Mormont ibid). Essentially, we 'know it when we see it'. In practice, for the rural at least, they must be realised by being fastened to real, and hence, usually, local conditions and resources to be fully experienced, articulated and reflected back. To explore 'rurality' as a social expression, therefore, the local has a role in providing context and expression. A full elaboration of the 'rural' becomes a type of dialectic between society and nature, with the 'local' the point of articulation – the go-between.

It seems we can suggest, then, that a cultural notion of the 'rural' does exists in society as a whole, and includes, is based upon, societies relationship with nature. We can also suggest that such a cultural relationship spans the more complex discourse of internal nature (civilisation, moral development, citizenship) and external nature, of materials and economy; which 'culture' can mediate. Such a mediation has obliged a localised reality to explore, but this localised reality is under threat. A further key question, therefore, becomes important – is there still or could there be a sustainable basis for 'the local' in society?

10.3 Practical Difficulty Number 2 : The Fate of the Local as a Development Concept

Among many others, Bryden (1994), Douglass (1984) and Smit and Brklacich (1989) suggest a key dimension of sustainability to be the condition of the more traditional idea of community, and the extent to which it is connected socially and economically to a local area. They face a general devaluing of the idea of 'community', however, among (rural) sociologists and even community development practitioners (see for example Greg Smith 1996). Community as a real, tangible phenomenon is even less in favour than is 'the local', particularly as means of identifying a 'rural condition'. Community is seen as the product of functionalist thinking if it is used to suggest a cohesive social entity, subordinating groups and individuals. Or it is seen as value-free but meaningless if used politically. The term does little to illuminate the actual condition of social relations underneath. Use of the term 'Local' has often been conflated with community, and criticised for the same reasons, (Murdoch and Day 1993) but is also under fire for other reasons – how meaningful is it as a spatial unit? Where does it start and finish? Do people feel local anymore?

Pahl (1966) is perhaps most noted for changing the academic fortunes of 'the rural local' as a concept. He concluded that *"any attempt to tie patterns of social relationships to specific geographical milieu is a singularly fruitless task"*. Nevertheless, the work of Stacey in 1969, has helped replace 'community' with the more acceptable idea of a 'local social system' - one which allowed the local to regain validity so long as it is not automatically equated with presumptions about quality of life, and the 'moral affluence' of the rural as opposed to the urban (Bradley and Lowe 1984 p6).

Work by social anthropologists such as Cohen (1985) re-addressed peoples sense of attachment to locality and their associated social meanings and experience, again by the recourse to language, values and beliefs employed to re-vitalise 'rural', through which people themselves depict the everyday events of their social world – the definition of locality is then centred on what Bradley and Lowe refer to as 'linguistic communities'. The key to this process is to discover the 'boundaries' implied by peoples expression of difference from others, which often invokes spatial/historical and familial associations of place.

Such a rendition of the local as 'place' though, even if set amidst wider processes of change, has been accused of implying an internal resistance and fixity within those people who promote 'place' locally, based on a recourse to the past, and to what Harvey (1989) terms (reactive) 'aesthetisation'. The reaction is against the idea of 'globalisation' and its images of technological facelessness.

Massey, for one, has challenged such an extreme view, (ibid.) but it cannot be denied that among the motivations driving the formation of community groups, there is likely to be a 'reactive' stance which is 'small versus big' (Gregg ibid.); resisting a new quarry or housing estate, for example, or seeking to arrest local economic decline. Such a reactive stance is inherent in the Rural Development Programme (see below) . I also highlighted the same role and influence of 'the aesthetic' Harvey treats so suspiciously as a valuable aspect of the local - as the 'cybernetic state' which at a 'system level' cultural empowerment as a strategy might help to facilitate. In summoning up 'the local', it is also impossible to avoid addressing 'the past'; an infamous source for conflict and division as much as joining people together.

However, the net result of a cultural empowerment process should be to furnish a positive engagement of people in a local sense with society at large, and not a retreat from it. Not least, the past, coupled to an appraisal of the present, and to a search into the future, has been shown to be a powerful means for galvanising a positive collective force for local change.

Such dimensions, of the local and the aesthetic, have a role, particularly in the production of rurality as an economic good, and, perhaps, as a means of apprehending sustainability. It may also be a counter, as I shall discuss, to the experience of social exclusion brought about by global change.

More recently, Massey (ibid.) and Cox and Mair (ibid.) have presented a substantial re-appraisal of the 'local' or of 'place'. Their approach sees the local as a nexus point for the convergence of social relations, with connectivities of varying spatial scale. In this way, the 'identities of place' become more provisional and open, but, crucially, can also be the basis for 'spontaneous organisation' (Allinson et al. 1993) around particular resources and opportunities.

This kind of depiction of place, amongst other things, makes more explicit the constitution of it in relation to its interactions with the 'outside'- (Massey ibid, p169), not only in market relations, but in the coming and going of 'outside interests' including commuters, tourists and recreational visitors. Such relations carry in the 'macro-social vision' of rurality which will then play a part in shaping the particular evolution of each local area. These relations emphasise the position of the local in a discourse triangle, indeed, at present, as dominated by it.

This vision of 'place', then, tends to depict it as a focal point for the convergence of essentially individualised paths, a manifestation of 'global' within trends to mobility. The key (physical) resources of place and their influences on these convergences are implied rather than traced out in this model in describing the basis for an emergence of place, yet must be intrinsic to the process. Cultural empowerment is a means to articulate these convergences and for those involved to become aware of them.

Callon (1986) and subsequently Murdoch (1994) have gone further and proposed more explicit models of 'place' which transcend divisions into separate social and natural entities, but invest natural ones with a role in the formation or dissolution of 'networks' of interest. Callons particular study combines research scientists, French scallop fishermen *and the scallops themselves* into such a network.

It is possible to describe rural development in similar terms, implicating local communities, visitors and other ecological and cultural objects as constituent parts in the formulation of localities, and in the discursive elaboration of the rural. Such a step back, which realises the dynamic interactivity of bio-geographic, other physical as well as 'social' actors in the renditions of place is a central part of a cultural empowerment strategy, providing for an increased presence of the local bio-geographic within rural discourse, and as suggested, empowering the ecological through the cultural.

Such an approach to describing 'local' – as a convergence of social and natural entities, allows openness; it describes 'local' as contingent and continuous networks of individual actors rather than communities, weaving in the role of the physical setting and its 'natural actors' as well as the usual social ones.

It is likely that any one place will retain some elements of fixity and tradition highlighted by Cohen (and Harvey) combined with the converging networks of interest suggested by Massey or Murdoch. Indeed, this mix of traditional and 'incomer' is closer to the norm than either alternative. A key element of rural and community development research must be to test the condition of each of these dimensions and how they might interact. The issue of sustainability suggests a balance to be required between the former, more stable form of place, and the latter, contingent and adaptive one; a balance which needs to be encouraged by the right blend of external signals, and where necessary, methods of intervention.

10.4 Practical Difficulty Number 3 : Globalisation, the Free Market and Social Exclusion

The rhetoric of sustainability often urges us to adopt a global perspective, a global consciousness, in our thinking about ourselves, our sense of belonging, and our actions. But, as with the potential of social representations, this essentially abstract state of belonging must be translated into concrete terms, so that while we may think globally, we 'act locally'.

Our 'sense of belonging' influences the sources from which we accumulate values, and hence obligations and responsibilities, key dimensions to our sense of identity. The idea is, presumably, that a shift in this sense of belonging and identity to a global perspective ought to enhance the capacity for individuals to cultivate an awareness and moral responsibility for things like equity, social justice and sustainability. But the driving forces of globalisation are competition, aided and abetted by Information Technology, fuelling a need for mobility and a capacity for rapid change. These latter responses to globalisation are not conducive to the former. Take five key examples.

10.4.1 'Globalisation' facilitates Individualisation – the competition for difference

Globalised distribution and sourcing of goods provides (at least initially) for increased choice of products from around the world. This increased choice facilitates individualisation – the quest to be different, or to service highly personalised wants. There is some competitive merit in seeking to provide for those wants.

The 'customising' of wants in this way by individuals is driven by the urge to self realisation and identity. If any 'Global consciousness' is achieved, it will, for many, be constituted out of a shared activity of consumption, on the competitive consumption of difference. This is literally true, for example, in the demand for new tourism destinations. Whilst this demand might be described as a search for new experiences, it is also subsequently a means to demonstrate difference to others. The race to be different, to find difference, means each untapped tourist destination is 'consumed' and there is an inevitable incremental shift toward homogeneity.

Other consumer products – cars, clothes, processed foods, technology, are capable of continuous (competitive) change – consumption here is still that of 'uniqueness' – of being first, and being first to experience or display ownership of the new product. Globalisation, by unlocking new resources of difference, facilitates or promotes a race toward individualisation of identity. This identity through consumption becomes a chief sense of common purpose across people as a whole.

10.4.2 'Globalisation' favours (is reflected in) economies of scale

A key strategy for competing in a global market means achieving economies of scale – particularly in more day-to-day products where competition is on price. Scale economics foster an increasingly fundamental detachment from place. The small shop is out-competed by the large supermarket. The small farm by the large agri-business. Increasing concentration of services to achieve larger 'catchment areas' fuels the need for consumers and service users to be mobile and for needs to be met outside the home locale. There are, of course, models of production which emphasise flexibility and which suggest networks of smaller businesses as a more effective competitive strategy, but this is a strategy to compete in the 'demand for difference' marketplace discussed earlier – itself a global one.

10.4.3 Information technology removes the local

Information technology, a main agent of globalisation, also 'removes the local' not just in terms of shrinking space and time, but also in removing the need for the local in social life and social identity, or for a physical 'local' or local setting which provides context for the production of discursive meaning. Instead, web sites take over as the cyberspace equivalent of the local – but an anonymous local. The Internet aids non-local social identity and individualisation. The sheer number of 'web sites' and discussion groups allows any sort of interest to be met. Communities of interest form on a non-local basis, but 'places' (discussion sites – 'forums' and the like) in which social exchanges take place are implied rather than experienced. Ultimately, they need a 'real' experience of place to make them work.

10.4.4 Globilisation fosters inequality.

As Massey (ibid.) has observed, 'globilisation' is not experienced equally or in the same positive direction by everyone. Maintaining social integration through mobility, travel, and consumption, access to information technology, and time on 'the net' depend upon income and circumstance. Global de-regulation for free trade threatens the security of work. The associated 'mobility' of capital and (some) people means what is here today can be gone tomorrow. It reduces local population stability as a basis for people developing expectations about a shared (local) future, and dilutes the presence of and confidence in the past; the particular ways by which the locality has emerged and the value that may be placed on this.

10.4.5 The Cultural Centrifuge

Commins (1996) and others highlight a 'political economy of signs' as another 'symptom' of globilisation – marketing and media activity (again, aided and abetted by IT) 'extracts' and detaches cultural images from place into universal forms – part of the demand for difference. Local culture is lost in the centrifugal force of global imagery.

Rather than enable conditions of connectiveness and equity between people, then, the economic changes associated with globilisation, and in particular, new technology, can also be depicted as fostering a concurrent process of Social Exclusion and alienation. The experience of Social Exclusion is also one of dispossession – of the means to sustain a recognised identity which therefore supports self esteem. Local forms of knowledge, or resources, are either superfluous or detached from place. Loss through either of these means loosing possession of a role, and hence a place in society.

Local identities as historically, and hence, locally cultured forms, then, are under threat from processes of globilisation. If there is a local dimension to society it is increasingly populated by those who find themselves excluded, immobile, and by association, impoverished. Where policies or interventions do attempt to utilise local cultural identification as a means to enhance community participation and esteem, they often do not encourage but in fact impose a presumed version of local identity drawn from society as a whole. (Ray 1996, Mack 1997) the 'Middle' and the 'Top' continue to dominate the 'Bottom'.

Some of the rhetoric associated with countering Social Exclusion coincides with that of rural development; calling for a fostering of organised efforts by 'the Top', to increase participation 'at the Bottom', in the control of resources and in regulative institutions. This is a position promoted by the Rio Earth Summit Local Agenda 21 initiative, for example.

In Northern Ireland and the Highlands and Islands, as with other parts of Europe, the impact of Rural Development as a policy initiative also supports this re-instatement of the local. Rural Development gained political momentum in the need for reform of the Common Agricultural Policy (reform was itself largely the result of international pressures), and as an attempt to counter rural poverty in the face of a reducing contribution from agriculture to a 'rural' economy. It has proceeded by placing an emphasis on 'community participation' in identifying problems, resources, and solutions to local development. It has progressed, therefore, by building up the number, capacity and competencies of community groups, and subsequently, structuring these into supportive networks. Alongside the emphasis on 'community' has been a focus on the local, with a strong trend in the orchestration of support interventions toward territorial or Area-Based Strategies and partnership with community groups.

The ostensive objective of the key agencies involved in the rural development process has been to 'help create a new vision to rural society' - to give it a voice and a means of articulating its potential; in other words, to reclaim a validity to a rural identity. Countering exclusion, though, has meant finding better ways to build broad and inclusive participation in local communities, and it is this need which has led to the sort of work which is described in the third part of this chapter; work which pays more attention to exploring cultural and arts-based methods of communication, of knowledge generation, and of vision building, within a local setting.

The 'rhetoric' of social inclusion is also about integration and citizenship. These are terms now prominent in EU social policy (see for example, the EU white papers 'European Social Policy: A Way Forward for the Union' and 'Growth, Competitiveness, Employment, The Challenges to the Way Forward into the 21st Century' (1994)). A key element of inclusion is that of tackling unemployment, and a main plank of doing so is to look for 'new needs and services' – new roles, in a modern economy. Finding a new role for the rural excluded seems to be complimentary to, if not integral to, the demand for rurality, and hence sustainability.

The fate of 'rurality', then, is tied up in the response to the drive in society at large to 'become global', in particular, in countering dispossession and disconnection, and fostering instead pathways to social inclusion. In this, fostering the local provides a key nexus within which inclusion might develop.

10.5 Practical Difficulty Number 4 : How Big is Local ?

Ostrum (1990) has brought together documentation and analysis of a number of instances in which a small scale 'Common Pool Resource' (CPR), such as an inshore fishery or a groundwater basin, have been self governed by 'communities' through local systems of collective action. She highlights a number of key factors in the emergence of such local systems. Of these, three in particular are of interest :

Firstly, the stability of local populations which was highlighted earlier, and hence the extent to which individuals in a given area share a past, and expect to share a future. Key to this experience is the extent to which local cultural identities are shared by and are important to the individuals living there, for example, that of crofting in Scotland, or belonging to rural townlands and parishes in Northern Ireland.

As I have pointed out, however, contemporary factors in society such as counter-urbanisation, 'globalisation', fragmentation and 'distantiation' challenge the conditions for sustainable local community identities as a cohesive reference for collective action.

The second theme is spatial, and includes two key issues; firstly, the degree of 'dependence' on a local area experienced by different individuals for economic, social or psychological needs as a crucial element informing attitudes toward and participation in local development. Secondly, the nature of spatial boundaries, geographic or administrative, recognised by individuals in a given area and implicated in the construction and reproduction of local identities.

The extent to which such spatial 'units of belonging' and local dependence coincide with patterns in the territorial distribution of natural resources is seen as central to the integrity of local CPR management. Not least, the extent to which natural resources are a component of local or individual identities - in terms of landscape, occupation, history and heritage. Locating the appropriate geographic scales at which economic, social and natural systems meaningfully combine has been proposed as a basis for the definition of 'Local Territorial Units' for Area Based Strategies of integrated development (Chiesa 1995). But do they combine? To what extent do they have to?

As an ecology is unlikely to coincide with the boundaries of any one individuals' property, nor even one particular community, and as it is unlikely to be used for one single purpose within the area, it must be articulated and treated as a CPR and in so doing, cast within a recognisable local identity and vision. The ingredients of local identity and collective vision are key to the extent to which cooperation and common purpose is demonstrated across individuals living in a recognised area, and is a basis on which to foster a partnership responsibility with local people for the management of an ecology. In turn, the cultural articulation of the local rural toward such a vision provides contexts and settings in which other dimensions of development become viewed - in so doing, it becomes part of the capacity-building process at grassroots level.

A third theme is the Institutional Environment and its history. Work by Ray (ibid) and Commins (1995) has detailed the role played by institutional arrangements in influencing or facilitating local identities and local autonomy. The Crofters' Commission, for example, is key to the crofting identity in the Highlands and Islands. EU LEADER programmes 1 and 2[3], designed for rural development, have created new territorial units and encouraged a use of histories and identities as marketing tools. But the extent to which the administrative boundaries of policy initiatives, or public bodies, which are formed by the need for targeting limited resources, actually reflect underlying units of belonging and hence, 'loyalties' to local development objectives, is less certain.

More problematic still is the lack of communication between administrative bodies and 'local' social systems, the different spatial responsibilities and the different scales of action, which rarely coincide. As Ostrum notes, spatial dominance over action can squash local initiative based on fragile factors of trust and informal equity.

[3] LEADER : Links Between Actions for the Development of the Rural Economy

A key element of research, then, which has been left to a large extent untouched as an explicit objective, is to identify the basis on which stable socio-cultural identities meet with institutional frames on the one hand, and natural resource systems on the other. It has been left untouched largely because of the problems highlighted earlier, in particular, the uncertainty of a meaningful 'local' domain.

10.6 The Elaboration of a Strategy of Cultural Empowerment : Fostering An Action Research Consensus Approach

Having looked at a theoretical perspective for viewing a cultural dimension to sustainable development, and some of the associated difficulties in operationalising it, I will now report on some initiatives which aim to provide more information on the extent of these difficulties and to provide techniques which might overcome them.

Work in the elaboration and employment of sense of place is being conducted in separate but complimentary initiatives in the Highlands and Islands of Scotland, and in the Mournes area of Northern Ireland. In addition, plans are on the drawing board to further develop tools and techniques through additional initiatives in Northern Ireland as part of the Rural Development Councils Action Research Programme addressing Rural Social Exclusion.

Two pieces of work currently underway set out to address a number of the issues raised above, including :

- Exploring means of establishing the size of 'area' or place to which different individuals feel they belong
- The current feeling of 'community' people have and what it means, or to what extent networks and local independence are more important
- The extent of participation or opportunities for inclusion these two dimensions engender, particularly, the discovery of new roles and pathways to identity
- The extent to which sense of place can furnish a collective and integrated approach to CPR Management

10.6.1 Research in the Highlands and Islands – Duthchas - A Sense of Place[4]

Work in the Highlands and Islands is being conducted under the heading 'Duthchas' which is the Gaelic word for 'Sense of Place'. The existence of such a word is itself an indication of the depth of hereditary relationship with the land and the sense of place it imparts which remains in this part of the world, not least through the tradition and struggle of crofting history. The project has set out with the overall aim of

> 'developing a mechanism for the sustainable development and long term stewardship of the natural heritage of an area, based upon a collective understanding of the values of that area, a shared vision for its future and an integrated approach to future management'.

The work seeks to engage and work closely with the people of an area in elaborating the nature of local associations with place, both in economic and cultural terms, but also to integrate the remits of appropriate development agencies. The combined understanding is intended to form the basis for an integrated approach to development which sustains cultural, economic and natural systems.

The approach addresses a number of the issues raised about 'size' of area. The first stage of the project, currently underway, is to combine criteria drawn from the various participating agencies to identify notional areas by the tradition of statistical indicators combined with GIS mapping techniques. The exercise will serve to illustrate the patterns in social, cultural and economic variables across the region. Convergence in these criteria will begin to suggest 'boundaries' for Common Pool Management. However a crucial next step will be to work on the ground with local people to discover their own 'areas', which experience suggests are likely to be much smaller, perhaps as small as the village or 'townland'. The third stage will consist of 'linking affinities', that is, drawing out the next level of association between people in their most local settings and the next most stable scale at which there is a sense of belonging. Such a sense is likely to be built from the scale at which people move around in their day-to-day activities, and at which they feel they have some 'control' or influence over events taking place.

---

[4] I am grateful to Vanessa Halhead SNH, a chief architect in the development of this area of work, for providing me with the background to the initiative and explaining the rationale. I hope I have done it justice. Further queries about this initiative can be directed to her at SNH, North-West Region, Inverness

At this stage it will be necessary to examine the original administrative boundaries of agencies for adjustments which can facilitate integration between the 'felt local' and administrative spaces.

Within defined local areas, a subsequent stage is the production of a collective statement of local character for an area, a formalised expression a representation of the area. Producing such a statement would seek to involve as many as possible who have an interest in the area, or whose lives and interests are integrally connected to it, in the identification of the values and issues which affect it. Producing the statement is essentially about the assisting a teasing out of existing understandings, perhaps re-vitalising them and building them into a collective sense of place.

The approach acknowledges, and seeks to accommodate, the varying social relations threading through even the most cohesive of localities, and the extent to which they can have influence over the direction of events which take place there. It does so by invoking a partnership process to define levels of association (or, in Masseys' terms, convergence) with 'place' as a physical arena. A key crystallising force is essentially cultural in nature, the expression of value and character for the area drawing together diverse sectors of interest into a common identification of place.

10.6.2 Research in the Mournes

Research in the Mourne Mountains seeks to provide a 'baseline' of sense of place and seeks to draw out a picture of 'place' through the eyes of the people who live and work there, and to examine this in relation to the formation of and participation in local community groups engaging with rural development. In this case an 'area' in which the work could take place was already suggested by the situation of the Mournes themselves. The Mournes are a well recognised feature of the Northern Ireland landscape, and attract a large number of visitors. Even so it is easy to distinguish a number of smaller areas in and around them reflecting circumstances of access, aspect, and 'religion'.

The religious and national affinities which have tended to divide Catholics and Protestants in Northern Ireland is well known, and whilst in practice this is a divide denied by either 'side', historic events have nevertheless placed them in different areas. Even where areas are shared, a look at the detailed movements of people of each tradition often suggests a subtle separation is maintained and the degree to which they inter-relate with one another can be quite small.

The historical definition of separate areas have been 'consolidated' to a degree by the singularity of the Catholic church and parish, which has formed an immediate and cohesive cultural space for Catholic communities, around which other activities such as the Gaelic football or hockey club are also based. The 'cohesiveness' of the Catholic parish and its associated community activities has enabled such communities to respond much more quickly to the 'bottom-up, community-based approach of Rural Development policy. A sense of place and of identity is coupled with experience in the organisation of community activities.

Much more work has been required among Protestants to overcome both a cultural tradition of individual independence which prevails within the Protestant identity and a less straight-forward basis for a sense of place. The Protestant religion is not one but a number of different churches - Presbyterian, Free Presbyterian, Church of Ireland, Methodist, and so on, whose parishes overlap. This has prevented the same cohesive sense of place Catholics might have as a basis on which to develop, unless by association with a local town or village.

Exploring sense of place in the Mournes through qualitative interviews, survey and focus group work has revealed a 'nested' and variable structure to exist set within the greater identity of the Mournes, and a variable set of ideas about what is 'rural'.

Given the common awareness of religious distributions within even the smaller area of the North Mournes, nevertheless the Mournes exert a powerful influence cutting across people in different localities. They act as a beacon to the homecomer, and have been a key player in orchestrating livelihoods such as stone cutting, farming, and the regional supply of fresh water, which provide occupations which cut across 'traditions'.

Farmers in these areas remain a separate community which yet also follows a 'nested' pattern of very local 'home patches' and more widely based networks within and across the area, formed from farming activities and patterns of land ownership which seek to combine poorer mountainland with better quality land further away.

The separation of farmers from the local community is not complete because most must have off-farm jobs to survive; nevertheless as farmers, they are often outside of the rural development community groups. This is a problem - if 'rural' is landscape and ecology, the participation of farmers in local development is essential.

Beyond all this, from the quite specific sense of home articulated by many as a townland or valley, to the local parish or village, the Mournes is the 'economic network' which re-creates Massey's convergence model of place. Most people in the area regularly travel within a 40km radius to obtain work, shopping and leisure. The area is seasonally awash with outsiders as visitors or tourists. globilisation and centralisation are playing their part. Nevertheless, these relations play out the sense of place in 'going out' and 'coming back' to 'home', or experiencing the seasonal invasion of others. This applies as much to 'incomers' from the city as to indigenous residents.

The representations of 'rural' which appear across this social landscape reflect and consolidate different lifestyles and settings as part of the sense of place of individuals – some see rural as 'remote' and hence, confined to the more marginal parts of the Mournes. Others saw it in terms of the (more affluent) village and areas of housing – the rest is 'countryside'.

The familial, cultural, and identity aspects of life, including identification with 'rural', were found to be still potent in defining an individuals situation in relation to the nested array of 'Mourne places', and were also observed, both here and elsewhere in Northern Ireland, to have a strong influence over the process and extent of formation in local community groups – a significant amount of initial discourse is devoted to defining and negotiating the 'territory of affinity' the group has emerged from – area boundaries become defined by the expected affinities of those living in the area, and decisions to participate in the group depend on this same sense of affinity across individual households. Equally, some areas or individuals may be discounted, or discount themselves, because they are not 'rural'.

Very specific local affinities are a latent but significant element of peoples lives in rural Northern Ireland, but these are often so clumped and limited in size as to represent a problem to achieving useful areas of cohesion for engagement with sustainable development. The use of 'cultural empowerment' becomes both important yet complex, in negotiating a collective vision across different traditions and historic lines.

This sort of baseline study has helped to inform the approach to building a 'cultural empowerment strategy' however. It suggests a building block approach. In common with the approach taken in the Highlands and Islands, initial awareness raising and capacity building needs to take place within cohesive 'locales' – which in Northern Ireland may be smaller than can be the case in the Highlands. The principle remains the same. From these, affinities and strategic linkages can be made to link up wider scales in which a sense of place still operates. In some parts of Northern Ireland this process is already evident, and it is bio-geographic features which are typically the key informing principle.

10.6.3 Vision Building Techniques and Participatory Action Research

Plans are in the design stage, as it were, within the RDC for the further development of tools and techniques to culturally empower rural communities and improve participation in the development process. These tools and techniques are focused under two main methodologies – vision-building and participatory action research. It is worth stating some of the principles underpinning these methodologies because they represent important departures from conventional social science practice.

To elaborate or unlock 'cultural' experience and associated knowledge requires a different paradigm to the positivism traditionally employed. One such paradigm is outlined by Guba (1990) or Shotter (1995) as 'constructivist'. Constructivists seek to facilitate the orchestration of both academic and lay discourse in the production of a common understanding. Such a process is not positivist in that it builds on subjective, rather than objective experiences of the world. However it may compliment positivist research as part of the pooling process between technical information about natural resources, say, and social relationships to them.

The combined subjective picture may not be readily generalisable, but as a methodology it is replicable and adaptive. It forms the basis of an approach termed 'Participatory Action Research' which is a cornerstone of the work within the RDC. Briefly, Participatory Action Research is characterised by the following :

- A combination of a 'Real-time' element of problem solving, consensus building, and adaptation, with a generation of knowledge useful to academic and policy interests in the longer term;
- An explicit involvement of people for whom the issue or problem to be researched has a direct impact, those who experience, indeed live the problems and issues at stake. this in partnership with research expertise;
- The constructivist precept that 'all knowledge is equal' in defining and addressing issues and problems and
- A holistic and cross-disciplinary approach.

The key to the process is facilitation (of the partnership) and evaluation. Evaluative instruments form the mechanism for registering the results and effects of the Action Research process. Their design, to be sensitive and wide ranging, is vital.

The approach to 'Vision Building' will employ arts-based activities to facilitate what might (after Paolo Friere) be termed an 'emancipatory process' (Friere 1979). Employing a variety of techniques including visual arts (photography, video and painting) and drama, coupled with critical studies and group facilitation, can raise the consciousness of participants and help them to 'situate' themselves socially and culturally. This process has the potential to re-calibrate the influence of 'the global' in relation to the local, and to culturally empower. It is in effect, a method of training and investigation from which visions for the future direction and role of the local community can emerge and feed into the action research process.

The combination sets out to build on the dimensions of the 'discursive triangle' outlined in the first part of the chapter, and to strength 'the local' as a key element in realising visions of rurality.

Three projects to apply the lessons learnt so far, and develop the principles of a cultural empowerment strategy are taking place during 1997/98.

Leader II. The RDC is working with the Centre for Rural Studies, Queens University, to design and implement an action research vision-building project for participatory planning. The initiative explicitly addresses the local/professional interface described in the discourse triangle. A number of 'communities' will be invited to participate in an exercise geared to enabling more effective engagement with local planning processes and decision-making, and to assessing the potential of the locality for the production of 'environmental goods'. The emphasis will not be solely on the local, but also on building the capacity within government agencies to engage with local communities on more equal terms. Building capacity in both local and government domains will begin from a cultural base, employing vision-building tools. The sorts of alliances and networks which emerge during the empowerment process will be monitored to provide insights to the 'scale' and nature of the local which is suggested - how might natural and social actors interact to produce, through their networks, underlying development spaces or the parameters of common pool resources? Artificial administrative boundaries will ignored in allowing such an 'organic' process to emerge.

The participatory planning process will recognise the limits to what community groups and local communities can be expected to do under the resources of voluntary action. Actions to be taken by government departments or agencies will be linked to ones which communities can do to work in partnership. An 'aesthetic responsibility' will be in the hands of local people and an attempt will be made to engage and validate local knowledge in so doing. Once again, the vision building process will help raise awareness of the area and of the role of local people in shaping its future. Building a broad participation in the process at local level will be as important an aspect of the methodology as the planning process itself.

'Articulation'. Working with the Arts Council for Northern Ireland, techniques of arts-based activities are to be specifically explored as a vehicle for raising awareness, facilitating discussion and exploration, generating knowledge and consensus, and defining action plans. Here, art is used as a process of investigation rather than solely for its products, however, the products will themselves reflect an engagement with 'the local' and 'the rural'. This sort of approach is vital for overcoming barriers to communication, however they may be formed. It will also enable individuals to address and define their situation as actors in both local and global terms, and 'unpack' the influences stemming from each. This sort of 'action research' provides for new and potentially powerful methods for exploring sense of place and the structures of influence operating on individuals, as a means toward further developing or modulating strategies of intervention for cultural empowerment.

Future Search. In Fermanagh, a county in Northern Ireland, the 'Topped mountain History Group', a community based group, is presently embarking on its own 'vision-building technique' employing its accumulated resources on the history and folklaw of the area alongside recent a recent economic and demographic survey to raise local awareness toward a collective and inclusive development initiative in the area. A key wish of the group is to address the encroachment of suburbia into its locality from the district town of Enniskillen, bringing with it incomers with a lack of connection to place the initiative is intended to introduce both existing and newcomers in the community to the cultural resources of the area.

Working in conjunction with SNIP (the Sustainable Northern Ireland Programme, the Local Agenda 21 body for Northern Ireland) a 'Future Search' conference is being planned. Future Search operates over 4 days, focusing in turn on the past, present and future of an area, and drawing together the values, aspirations and needs so defined to agree a local action plan.

10.7 Drawing together the Strands : Distinguishing a Strategy of Cultural Empowerment in Sustainable Rural Development

This chapter has outlined the theoretical issues and provided some operational and practical signposts, for a cultural strategy for sustainable rural development described as 'cultural empowerment'. I will conclude by drawing together these various strands to distinguish the role and place of such a strategy within rural development policy and practice.

I began by defining and describing 'culture' in discursive terms. In discursive terms, culture becomes based around ways in which knowledge and experience is formed and articulated, on the process by which information and meaning is socially constructed and exchanged, revealing a model of culture which has 'cybernetic' potential – i.e., to reach and maintain an internally recognised complex state of equilibrium. The cultural route offers the means to develop a sense of sustainability within social life; an aesthetics of (or for) sustainability.

Culture, then, in its discursive sense, is concerned with 'knowing the world'. This gives us a clue as to the nature of a strategy of intervention designed to enhance this; that it needs also to address active and participatory learning. The principle of participation in the generation of knowledge across (local) people should underpin a strategy for 'cultural empowerment'. Such a strategy should seek to provide the avenues by which local people can become involved in the articulation and interpretation of experience and the generation of knowledge, and in so doing, for them to be contributory to the development process as a whole in more equitable terms. Cultural empowerment is also about making those equitable terms comfortable, familiar; continual and sustainable, as a basis for citizenship.

What makes this different from a policy of improving education and training as a core element of development? Or of awareness-raising on sustainability issues through the media?

Firstly, whilst the description given of culture as discursive sought to expand its meaning beyond associations with arts, crafts and festivals in rural development, nevertheless these are not rejected in this expanded meaning, indeed on the contrary; they should be recognised as key tools. 'Art' can be used as an investigative tool, as a part of an action research methodology, and as a means of critical assessment and emancipation. Cultural empowerment seeks to employ a variety of means of experience and understanding to furnish a participatory learning process.

Secondly, Cultural empowerment was introduced as a strategy based on an engagement with local social, cultural and physical characteristics, and linking where possible the past, present and future together to begin to define and realise a locality, and the convergence of people and their interests which have produced it. By so doing, the presence of 'the local' in the cybernetic process of discursive change is enhanced.

The rural local includes, most importantly, the bio-geographic and ecological. As such, a cultural empowerment strategy seeks also to 'give voice' or more realistically, discursive presence, to natural systems or their constituent parts, as elements in the network of people and things which interact to produce rural localities. An outcome of this enhanced interchange (or dialectic) between local socio-cultural and ecological domains is the production of a rural 'aesthetic' which captures and 'transmits' a condition of sustainability holistically.

An emphasis on the local, and also on participation, inevitably links such a strategy with the process of 'community development' already pursued within rural development policy and practice. Both are concerned with capacity building and 'empowerment' - instilling a level of autonomous self sufficiency at the local level in what have been centralised activities. Again, why suggest an alternative term; 'cultural empowerment' when community development might do?

Community development, in Northern Ireland at least, has become strongly associated with achieving economic goals or with the capacity of a local population to compete for public resources. Because it is abstract and indirect, a cultural strategy of empowerment has been largely ignored or misunderstood as too 'soft' or unhelpful to the greater imperative of economic development in rural areas. Early, exploratory work in community development recognised the role of culture and social development but could not grasp and demonstrate its worth in the sorts of terms and time scales required by funding agencies. Economic goals were felt more concrete and more easily measured, but in fact, as a focus for community development, they have so far fared no better. The demands made on what is voluntary effort, to conceive, plan and manage a sustainable (and substantial) economic project, have begun to prove too much.

There is an understanding beginning to form amongst statutory agencies that the goals and meaning of community development may need reviewed, toward more realistic, (and more social or cultural) goals which link to or compliment entrepreneurial activity in local areas. The role conceived for community development can be re-cast to one of servicing a 'development of the local', rather than of 'local development', of developing a sense of place and of common resources in which local (globally informed) development takes place.

Cultural empowerment in the manner outlined here, of participatory learning and discourse, using 'the local' as a main point of reference, becomes the key informing principle for this form of community development, and distinguishes it from other forms of capacity building at local level which have an economic or political objective.

Current forms of community development for economic development have also tended to make the assumptions about the existence of cohesive 'communities' discussed earlier and found wanting. It is important to avoid assuming that either a strategy of cultural empowerment or of community development can produce a 'community' in the traditional sense – it would be a mistake to imagine such communities are possible (or ever were possible).

A cultural empowerment strategy for community development does not seek 'community' as an outcome; rather, its goal is an enhancement of local discourse between individuals toward collectively informed actions, actions informed by a mutual exploration of place. The strategy assists what might be described as a community development process to the extent of raising a collective awareness and consensus about local characteristics and their value. Vision-building tools, and participatory learning through action research, are means toward this goal.

A collective awareness and consensus must be adaptable, though, if it is to service sustainability; it must be about 'process' as much as 'product', and so the extent of a 'development of community' in the strategy must go further, to install the capacity and desire for an ongoing level of discursive activity within and about the local domain. Capacity includes places for discursive activity, means of contribution, appreciation, and elaboration. It includes formal and informal exchange, arts and debate, festivals and stewardship. Desire includes accessibility, meaningful activity and sense of participation. Desire also means ensuring a sense of legitimacy to participation in this sort of discursive activity.

10.7.1 The Issue of Legitimacy

A cultural strategy for the local must be backed up by an atmosphere of legitimacy if it is to work. Two aspects of legitimacy need to be addressed.

Firstly, better definition of the role of local citizens in development, respected by policy and practice. This means identifying an appropriate devolution of responsibility to the local level. However, too much responsibility goes beyond what can reasonably be expected of the voluntary contributions of local citizens. There is already a suspicion that in community development, the State is off-loading its own responsibilities. Responsibility also creates institutional structures for accountability which tend to exclude the majority of local citizens and inhibit many others.

To get the balance of responsibility between the local and the State right, a cultural strategy should not only rest at the local level, but should also be applied and recognised within the 'Top'; the academic and professional discourse domain of the discourse triangle. This is essentially about inserting (recognition of) the local into the 'Top' as a legitimate discourse domain in its own right. This recognition and understanding is needed to enhance the conditions for partnership between local and non local actors in pursuing different tasks and goals in rural development.

Exploring the establishment of these forms of legitimacy in real terms is a goal of the LEADER II programme for participatory planning described earlier. The principle highlights an important fact; within the discourse triangle, the legitimacy of an individual discourse domain depends on its ability to be recognised and to inform other domains; the legitimacy of 'the local' is as important as the legitimacy of local actors.

Secondly, this principle, of the legitimacy of the local domain, applies not only to statutory agencies, but equally to what individuals in society as a whole consider to be legitimate for the investment of their activity and thinking, their identity (identification) and sense of belonging. A consumer identity as the main sense of collective activity in society may be a key response to globalisation, for example, one which tends to discard or marginalise the local.

To re-build or coalesce 'the local' in larger society depends upon the legitimacy people perceive in investing in local life. An individuals' investments; of time, and of economic, social and cultural capital, and the decisions made to cooperate or compete, to participate in local development, or a development of the local, are informed by the pattern of legitimacies (or norms) within society at any one time - held as a kind of collective consciousness[5]. It is the subtlety with which options are accepted or rejected over time and across individuals which maintains the pattern of legitimacy – a pattern which is the cultural 'product' of societal discourse in the 'Middle' domain of the discourse triangle.

The collective conscious of society today is suffused and informed by the belief in a life becoming globalised. This message is pervasive within the discourse of wider society, and of the 'Top', and so must also affect the extent, nature, and viability of a local domain of discourse. Once again, vision building and critical reflection, through new tools and techniques, are exercises providing the opportunity to assess the legitimacy of 'global' and of 'local' and the balance between the two. The realities of a society moving between the local and the 'global' cannot be ignored, yet concrete opportunities for citizenship, for cultural participation and for shaping local life, may provide counterpoints to detachment and individuality.

Reconciling the 'global' cultural zeitgeist in society with the 'rural local' means 're-inserting' the local back into this societal discourse domain just as its insertion is required in the 'Top', strengthening its legitimacy and therefore, the legitimacy of choosing to participate in it.

A 'cultural empowerment strategy' must operate with all three of the discourse domains of the triangle in mind. It cannot focus only on 'developing the local' or a sense of community, if that is done to the isolation or neglect of the position of the local in larger, globalising society. It cannot assume such a strategy will be understood and appreciated by professionals and policy-makers if they too are not addressed, indeed, exposed to similar exercises of reflection. A dialectic is required between the three domains utilising a cultural currency, using the 'local' as a key aesthetic reference.

10.7.2 Postscript

---

[5] It is this pattern, and the collective sense of it held within society that is the target of a cybernetic cultural model.

In seeking to articulate a sense of a ceaseless mobility in present society, and its effects on perceptions of the world, Thrift (1993) draws attention to an 'aesthetics of disappearance' which can be recognised as; 'the landscape (becoming) a blur, a streak viewed from a moving platform' (p221)... from the window of a moving car. The task of a cultural empowerment of the local is to seek to demonstrate 'places'. To paraphrase Thrift; 'every journey includes stops in places; continuity requires discontinuity'. The role of the rural local may be to provide those places, to incorporate the ecological into those places, and to provide the opportunities to stop and reflect.

The realisation of an abstract notion of rurality needs concrete places or locales. The insertion of a rural locality into modern, 'global' society is an insertion of pauses into an increasingly frenetic mobility of people through space and time; the formation of oases in speed.

References

Allinson P, Murdoch J, Garrod G, Lowe P (1993) A Sustainable Rural Economy ?

Paper presented at the Agricultural Economics Society conference 'Problems and Prospects for a Sustainable Rural Economy'; London 1993 February.

Bourdieu Pierre *Outline of a Theory of Practice*. Cambridge University Press 1977.

Bradley Tony and Lowe,Phillip. Introduction: Locality, Rurality and Social Theory. In; *Locality and Rurality: Economy and Society in Rural Regions*. Bradley,Tony and Lowe Phillip, eds. Norwich, Geo Books, 1984.

Bryden John. 'Towards Sustainable Rural Communities – From Theory to Action'. In *Towards Sustainable Rural Communities: The Guelph Seminar Series*, Canada, University of Guelph Publications, 1994

Callon Michael. 'Some Elements of a Sociology of Translation: Domestication of Scallpops and the Fishermen of St Brieucs Bay'. In *Power, Action, Belief-A New Sociology of Knowledge ?* Law John, ed. London, Routledge, 1986.

Chiesa R. Sustainable Rural Development : Concept, Policy and Practice. Working Paper, Rural Social Science Technical Cooperation Network, Teagac, Dublin, 1995.

Clifford S, King A. New Milestones : Sculpture, Community And The Land. The Green Book 1989; 3 (1): 40-4

Commins,P NESC Report. Teagasc, Dublin, 1995.

Commins P. Social Exclusion In Rural Europe. Paper Given At The Conference Rural Social Exclusion, 1996 November, Cork, Republic of Ireland

Cohen,Anthony. *The Symbolic Construction of Community*. London, Routledge. 1985

Conway G.R. The Properties of Agro-Ecosystems. *Agricultural Systems* 1987; *24: 96–117*

Cooke Phillip. *Back to the Future: Modernity, Postmodernity and Locality*. London, Unwin and Hyman, 1990.

Cox KR, Mair A. From Localised Social Structures to Localities as Agents. *Environment and Planning* A 1991; 23: 197-213

Douglass Gordon. 'The Meanings of Agricultural Sustainability'. In *Agicultural Sustainability in a Changing World Order*. Douglass,Gordon, ed. Boulder, Colorado, Westview Press, 1984.

European Commission. *Growth, Competitiveness, Employment : The Challenges and Ways forward into the 21st Centuary*. Luxemburg, European Commission, 1994

European Commission. *European Social Policy: A Way Forward for the Union*. Luxemburg, European Commission, 1994

Friere Paulo. *Pedagogy of the Oppressed*. London: Sheed and Ward, 1979.

Gregory David. *Geographical Imaginations*. Oxford, Blackwell, 1994.

Guba Edward.G. *The Paradigm Dialogue*. Sage Publications, 1990.

Harvey David. *The Condition of Postmodernity*. Oxford, Blackwell, 1989

Hoggart K. Lets' Do Away With Rural. Journal of Rural Studies 1990; 6 (3) 1990

Halfacree K. Locality and Social Representation: Space, Discourse and Alternative Definitions of the Rural. Journal of Rural Studies 1993; 9 (1): 23-37

Jones O. Lay Discourses of the Rural: Developments and Implications for Rural Studies. Journal of Rural Studies. 1995; 11 (1): 35-49

Mack N Seeking Local Citizenship-Toward a New Sense of the Local ? Paper Presented to Working Group 3 : The Social Construction of Locality in the Post-Modern Countryside. XVII Congress of the European Society for Rural Sociology
1997 25-29 August, Chania, Greece.

Massey Doreen. *Space, Place and Gender*. Cambridge, Cambridge Polity Press, 1994.

Milton Kay. *Environmentalism And Cultural Theory-Exploring The Role Of Anthropology In Environmental Discourse*. London, Routledge, 1996

Mormont Marc. 'Who is Rural ? or, How to be Rural: towards a Sociology of the Rural' In *Rural Restructuring: Global Processes and Their Responses*. Terry Marsden, Phillip Lowe, Sarah Whatmore, eds. London, David Fulton, 1990.

Murdoch Jonathan. *Weaving the Seamless Web: A Consideration of Network Analysis and its Potential Application to the Study of the Rural Economy*. Working Paper 3, Centre for Rural Economy, Newcastle, University of Newcastle upon Tyne. 1994

Moscovici Serge. 'The Phenomenon of Social Representations' In *Social Representations* Farr,R. and Moscovici,S eds. Cambridge, Cambridge University Press, 1984.

Noorgard R. Co-evolutionary Development Potential. *Journal of Land Economics;* 1984 *: 60 (2) : 160-173.*

Ostrum Elinor. *Governing the Commons: The Evolution of Institutions for Collective Action*. Cambridge, Cambridge University Press, 1990.

Pahl RE. The Rural-Urban Continuum. Sociologia Ruralis 1966; 6: 299-327

Ray Chris. Local Rural Development Initiatives in Western Europe: Psychological and Ideological Perpsectives. *Paper presented at the Agricultural Economics Society Postgraduate Conference, 1996, February. Aberystwyth.*

Shotter John. *Cultural Politics of Everyday Life: Social Constructivism, Rhetoric and Knowing of the Third Kind*. Buckingham, Open University Press, 1995

Smit B, Brklacich M. Sustainable Development and the Analysis of Rural Systems. Journal of Rural Studies 1989; 5 (4): 405-414.

Smith G. Ties Nets and an Elastic Bund: Community in the Postmodern City. Community Development Journal 1996; 31 (30) : 250-259

Smith Neil. *Uneven Development*. Oxford, Blackwell, 1990.

Stacey M. The Myth of Community Studies. Brit J Soc 1969; 20 (2): 34-47.

11 QUALITATIVE GROWTH – EMPLOYMENT : *A Revisited Relationship*

Laurent Grimal and Charilaos Kephaliacos

Introduction.

The current economic development is characterized by two fundamental imbalances: an under-exploitation of labor force resources and an over-exploitation of natural resources and energy. This is certainly the reason why an important theoretical debate focuses on the relationships between the different measures directed to the environmental protection and the rate of employment. This debate follows, in general, two separated directions :

- According to the first, environmental protection policies imposed on firms represent supplementary costs; these costs drive an increase in job destruction.
- According to the second direction, environmental protection activities are considered as a constantly developing economic sector and an important potential for job creation.

However, most of the studies adopting either of these two directions do not undertake a dynamic approach to the job-environment connection. Yet the concept of sustainable development, defined in the Bruntland Commission's report, as "*a development which meets the needs of the present generation without compromising the ability of future generations to meet their own needs*" (World Commission on Environment and Development, 1988 , p. 51), asks for a long term and dynamic approach.

Even if many authors refer explicitly to the sustainable development concept, they do not use a clear operational definition of this concept. J. Pezzey. (1992)

counts, in the literature, thirty-seven different meanings of the sustainable development. S. Faucheux et al. (1993), classify these definitions following two main components: on the one hand, the degree of substitutability between different resources, especially between natural capital and physical capital and on the other hand, the assumption of hypothesis of rationality of agents an either substantial rationality or procedural rationality.

The aim of this article is to look for the coherence, in a dynamic perspective, between the objective of sustainable development and the objective of job creation in activities related to environmental protection.

In fact, a sustainable development path implies structural changes, both in the system of production and in the system of social and individual preferences. Such structural changes appear as modifications in the processes of production and, in a complementary manner, in the utilization of human capital. They also take the form of change in the weight allocated by public authorities to environmental protection policies.

Observation of the recent evolution of environmental activities in France and Germany reveals significant differences in the paths adopted by these two countries. These differences will have consequences, in the long-term, on the structure, the number and the viability of jobs in the area of environmental activities.

In the first part of this study we adopt an operational concept stemming from the sustainable development notion: qualitative growth (Renn, 1995).

The second part presents some statistics on environmental jobs and activities in France and in Germany.

In the third part we try to reformulate the jobs-environment relationship while taking account of the qualitative growth criterion. To this end, we propose a distinction between environmental activities. We show then that the development of a category of these activities (and the jobs associated) is not consistent with the qualitative growth criterion.

11.1 Qualitative growth : an operational concept of sustainable development.

Economic growth is commonly expressed in quantitative terms as an increase of GDP. In that way, it does not matter how this growth has been obtained.

Damages inflicted on the natural environment are not taken into account; worse, when the GDP is calculated, it includes pollution abatement activities[1].

The explanation of this situation resides in the central idea of neoclassical economics that market mechanisms play a self-regulatory role. There is an optimal path of resources exploitation: as a natural resource is consumed, its price increases and economic agents will have an incentive either to find a substitute resource or to adopt another production process. The diminishing of natural resources is then compensated by an increase of newly created resources and by the adoption of innovations. This idea is summarized by Hartwick's rule stipulating that "*all the rents issued of intertemporal optimal management of the exhaustion of natural resources are invested in the reproduced capital that will have to substitute to the natural resources used in the production*" (Hartwick 1977). In the long term, the economy tends to a path of balanced growth (See, for example, Solow 1974).

As the opposite of this form of growth, many authors have proposed different ways to take into account the damage inflicted on the environment by economic activity. They all indicate the necessity to modify the market mechanisms in order to favor the environment and to evaluate the extent to which economic sphere can be subordinated to the aim of the environmental protection.

Some authors propose, on the grounds of the second law of thermodynamics, the adoption of a negative growth (Georgescu and Roegen 1979). Others maintain that if the economies integrate truly ecological preoccupation, they have to turn towards the stationary state (Daly 1991). Finally, R. Passet (1994) by adding further the negentropy phenomenon, suggests an improvement in the economy's prospects of growth while reducing the negative interactions between ecological and economic spheres.

Whatever the theoretical or ideological orientation followed by the different authors, a consensus exists as to the necessity to reduce environmental pollution and generally to orient economies towards sustainable development. Thus, it is increasingly urgent for economic policy to rely on an operational concept that allows measures in this direction to be taken. Such a concept has recently been proposed by O. Renn (1995): qualitative growth.

Qualitative growth can become a preliminary to the sustainable development while giving it an operational content from the economic point of view.

---

[1] Much work has been done on incorporating environmental problems in GDP accounts, either calculating a 'green GPD' or creating 'satellite accounts'. This last solution has been held back in France, with the creation of an Interministerial Commission of natural patrimony accounts in 1978 (Comolet and Weber 1990; see also Pearce, 1995).

11.1.1 How does one define qualitative growth ?

The first idea underlying the qualitative growth process is the will to decrease the strong complementarity between natural capital[2] and the other – human and technical – resources of the economy. A characteristic example of this process is given by Germany where, for the period 1980 – 1990, the GDP increased by 23% while the level of polluting emissions (especially atmospheric pollution) and the emission of wastes have decreased (OECD, 1993 a, p. 42).

Qualitative growth is a process during which the resource productivity (natural capital, human capital and technical capital) increases. But the characteristic of this type of growth is that the improvement of the population's wellbeing also comes from the reduction of damages to the environment.

We can simply express the qualitative growth process in the following manner. Let Q = f(E, L, K) be a production function where E is the natural capital, L the human capital and K the technical capital. A qualitative growth (positive) is characterized by a simultaneous increase of the mean productivity of all production factors (Q / E, Q / L and Q / K) subject to the supplementary condition that the increase of the ratio Q/E is due to the absolute decrease of the denominator.

- The increase of this last ratio is in general technically and economically very difficult due to of a high degree of complementarity between natural resources and the other resources. Thus the qualitative growth rests, among other things, on the improvement in the long term of the substitutability between these two types of resources.

However, in order to attain this objective the economy has to impose strict limits[3] on the exploitation and management of natural resources, notably:

- the exploitation of renewable natural resources (energy and raw materials) can only be used at a rate which allows a balance to come about between consumption and regeneration;

---

[2] See Berkes and Folkes (1992, p. 2): "Natural capital consists of three major components : (1) non-renewable resources that are extracted from ecosystems, (2) renewable resources that are produced and maintained by the processes and functions of ecosystems, (3) environmental services such as maintenance of the quality of the atmosphere, climate,..., waste assimilation, recycling of nutrients, generation of soils, and the maintenance of a vast genetic library. These crucial services are generated and sustained by the work of ecosystems".

[3] We are not interested here in the real difficulties associated with the implementation of these limits in the economy.

- that non-renewable natural resources (energy and raw materials) are exploited at a rate that allows their progressive replacement either by renewable natural resources or by other newly created resources;
- that emissions of wastes are inferior to the assimilation capacities of the ecosystems. Note that human intervention to increase these capacities will consequently be limited in the future.

These rules of management of the natural capital, besides the fact that they require treatment of the ecological problems at the source, oblige us to lean on the fundamental link between natural resources and labor. In fact, limiting the use of natural resources means consequently a reduction, at least partly, of the labor force used in the processing of pollution. The qualitative growth does not authorize a process of resource transformation resulting in a quantity of pollution so important that it is necessary to reprocess it thereafter.

In view to deepen the jobs-environment relationship we are thus led to effectuate, inside the environmental sector, a distinction between the activities intervening upstream in the process of production (*innovation and use of cleaner production technologies*) and those intervening downstream (*end-of-pipe technologies*) [see 11.3.1.1 below].

- Concerning the two other ratios, one cannot generally state if their increase comes from a stronger increase in the numerator as compared to that in the denominator or from a decrease in the denominator, as the productivity of the labor factor, for example, increases because of addition of new knowledge or immaterial services in the production process.

In order to investigate the functioning of the qualitative growth process it is then necessary to decompose the human capital into skilled and unskilled labor and the physical capital into technical equipment and immaterial capital (know how, learning by doing and other sources of non-decreasing scale returns [see 11.3.2.1 below].

Thus, a qualitative growth orientation of the economy will imply a decrease of manual labor in favour of skilled labor. Similarly, the proportion of immaterial capital implied in the production process will increase relatively to the technical equipment. This evolution will therefore have consequences on the jobs structure and on income distribution.

In summary, qualitative growth can be characterized by the simultaneous increase of production factors' productivity accompanied by a decrease in the use of natural resources. What kind of forces or mechanisms can push an economy to engage in such a growth? Can the market mechanisms be attributed some autonomous role in that direction?

11.1.2. What is the interest of such a growth ?

Environmental protection policies[4] are often considered to hamper the competitiveness of firms. For many authors environmental protection, especially if it is implemented via 'Command and Control' type measures, is responsible for firms' production costs increases (See, for example, Jaffe et al., 1995; Barbera and McConnel 1990). They influence the costs of compliance with environmental norms which are translated into an increase in pollution abatement expenses.

Furthermore, these policies are considered an impediment to foreign investments in countries where environmental norms are viewed too strictly. It is also observed that 'dirty' industries, initially operating in these countries, are tempted to delocalize to countries where norms are more lax (Tobey 1990, Blazejczak 1993).

In face of these analyses, we remark, in the first place, that the weight of costs linked to environmental protection are marginal in comparison to total production costs. They represent on the average only 1 to 2 per cent, according to industrial sources (Jaffe et al. 1995).

Table 11.1 shows, the share of pollution abatement costs in total production costs of German industry is rather weak, even for heavily polluting industries. Yet, in this country, emission norms are relatively strict compared to most European countries.

Secondly, empirical studies (Tobey 1990) show that firms do not proceed generally to reallocations of their activities in order to avoid an increase of production costs due to environmental constraints. In fact, most observed movements in international investment can be imputed to better market opportunities (growth in the long term) in host countries, rather than to a increased level of environmental norms in the originating home countries.

Finally, estimations show that in the United States, between 1971 and 1983, only 155 firms of the industrial sector went bankrupt – representing a loss of 32 900 jobs due to environmental protection policies (*Données Economiques de l'Environnement, 1992–1993*, 1993, p. 87). In France, analogous estimations

[4] As well as norms and regulations which at the same time regulate emissions and processes of production, environmental protection policies can use economic instruments: taxes, subsidies, licences, etc.
For a survey on the relationships between environmental protection policies and competitiveness from a macro-economic point of view, see Medhurst (1993).

indicate only some hundred losses of jobs for the same period. Most of the establishments that have closed were bound to disappear, by reason of obsolete production techniques. In fact, the strengthening of norms has accelerated an inescapable tendency.

Table 11.1 Pollution abatement costs as a percentage of total production costs.

Industry	*Germany (1989)*
Food, beverage and tobacco	0.35
Textiles and leather	0.26
Wood and wood products	0.46
Pulp and paper	1.2
Chemicals	1.7
Iron and steel	0.84
Machinery	0.28

Source : Tobey, Smets 1996, p. 69.

Apart from the case of job losses due to firm bankruptcy, another less drastic consequence of environmental policies is the job loss due to a reduction in the activity level. For example, in Germany, between 1980 and 1990, a loss of 55 000 jobs has been observed because of the increase of production costs. This increase was necessary for the compliance of industrial installations with environmental regulation (Kleinschmidt et al. 1994). Consequently, only a marginal place can be attributed to environmental protection policies in the increase of the unemployment experienced in recent years by the western economies. However, the preceding analyses view the impacts of environmental policies in a static framework. In this framework, technology – the processes of production and product variety – but also demand are fixed. Stricter environmental measures inevitably provoke increased costs. If we include some induced effects in the analysis, it is possible to sustain that these measures lead in the longer term to an improvement of the whole competitiveness of firms. Even more, they can induce the emergency of a economic *merchant* sector linked to the environmental protection.

A first induced effect of the environmental policy is to progressively incite firms to decrease pollution emissions in order to diminish their total production costs. This behavior shift, on a micro-economic level, consists in a radically new perception of the relationship between firms' competitiveness and environmental constraints. In general, polluting rejects are analyzed in economic theory as negative externalities. Firms are invited to internalize them. According to the economists M. Porter and C. Van Der Linde (1995), pollution is the sign of an incomplete and rather bad utilization of resources in the process of production. In fact, to pollute it using the different inputs in an inefficient manner. The reason for this inefficiency resides in the inappropriateness of the production technique used. Consequently, to be in conformity with environmental regulations, or even to anticipate them by innovating in the right direction means to improve the resource productivity used in the process of production. This improvement results thereafter in a cost reduction and therefore in competitiveness gains

More specifically, this adjustment can be translated into a decrease of pollution abatement costs, but also in a reduction of input quantities used, particularly of natural resources and energy. A good estimate of the evolution towards an improvement in the resources productivity is the energy-efficiency indicator defined as the energy over production ratio. In Germany, for example, the energy-efficiency decreased from an index base 100 in 1985 to 87 by 1990. For France, this indicator has decreased, for the same period, from 100 to 91 (see Thomas and Lavoux 1995).

Thanks to learning by doing and learning by interacting, firms can obtain greater benefits by using the existent processes. But even more, firms can expand internal R&D activities in order to develop production processes with less use of the environment as a wastes receptacle. This 'ecological modernization' then takes the form of an offensive innovation strategy so as to induce modifications of the market conditions (Heraud and Llerena 1992) serving the firm's interest. In this way, innovative firms are more able to earn new competitive advantages and additional market shares. Thereafter, these firms will be able to sell their production processes and equipment.

Thus, according to these two authors, firms should not interpret environmental protection measures as additional constraints reducing their competitiveness, but rather as a necessity pushing them to improve their production techniques and the quality of their products, thus increasing up their competitive advantages.

It turns out then that, if firms combat pollution at its source by looking for new production processes, this is not only to decrease ecological damages, but also to improve the productive efficiency and consequently the competitiveness of firms. Even if these efforts do not inevitably imply net creation of jobs, nevertheless they constitute a necessary condition for this creation.

A second effect induced by environmental policies is the emergence of a sector linked to environmental protection. One can often find a strong correlation between the level of environmental regulations in a country and the emergence and growth of its environmental sector (Jacobs 1994). This sector is a response to the firms' demand that seek to decrease their polluting wastes by introducing not only new abatement pollution activities but also cleaner production processes.

The emergence of this sector is currently strengthened by the appearance of a strong demand for the products respecting the environment. Thus, many inquires show a growing sensitivity of the consumers to the environmental protection. According to a poll undertaken in France by CREDOC in 1994, 43.2% of questioned persons declared themselves willing to pay 10% more for products preserving the environment. Even more, 66.6% of them were willing always to pay 10% more for products recognized as less to health harmful (see Dobre and Hammer 1995).

Firms are driven to take into account this autonomous demand evolution for products preserving the environment. Producing 'eco-products'[5] helps them earn new market shares. In order to be in phase with this demand evolution, firms are led to adopt adequate marketing policies (driven to the eco-products consumer culture). This marketing strategy has to be accompanied by an innovation strategy using R&D towards cleaner production processes.

It appears that, when an initial impulse in favor of the environment has been given by the public authorities, firms, in their own interest, are susceptible to reduce their pressure on environment not only by innovating at the end of the pipe but also by integrating cleaner technical processes. This dynamic phenomenon, illustrating some inherent properties of capitalist markets, seems to lead in a relatively autonomous manner to an improvement of the environmental quality and permit the orientation of society towards qualitative growth.

It is the consideration of these new elements that leads us to investigate in a new perspective the jobs-environment relationship (Section 11.3). However, it is necessary to present, before that an inventory of the whole set of jobs and the economic activities linked to the environmental protection in France as well as in Germany.

11.2 Environmental jobs and activities.

On the one hand, environmental policies can provoke few losses of jobs. But, on the other hand, they can give a strong impulse to the development of activities

[5] An eco-product is defined as 'a consumer good emerged since five years at the most and which the use or the scrap is less polluting that products make the same service' (JULIA 1994, p. 121). Note that the manner in which the eco-product is made, does not enter the game for its classification. A product stemming from a polluting production process can be classified among eco-products.

related to environmental protection. Thus, an increase of environmental jobs can appear.

When studying the relationship between environmental protection and employment, we can distinguish the *'environmental jobs'* and the *'professions linked to the environment'*. Environmental jobs are practiced in structures which the main activity is the protection of the environment. The professions linked to the environment square with *'an environmental practice at the level of the station work'* (see Roch 1996, p. 13).

In this section, we present first *the environmental jobs* in France and Germany and, second, some indications about the structure of the environmental activities in the two countries.

11.2.1 The categories of environmental jobs in France and Germany.

Environmental jobs 'are jobs that have been created in the design to take into account the environment in the economic activities in general (merchant or non merchant)' (Roch 1996, p. 19).

Statistically, a decomposition is operated in the account of these jobs. We distinguish between :

- Direct jobs. These are jobs, in the manufacture or in the services sectors, occupied in full or part time in the environment area.
- Indirect jobs. These are jobs in sectors (manufacture or services) furnishing intermediate goods and services to firms engaged in the environmental activity.

Globally speaking, the number of jobs has a tendency to grow:

- In France, they have increased by 1% in 1991 and 0.9% in 1992. Since 1987, about 4 000 supplementary jobs have been created annually in the environmental sector.
- In Germany (in Länder of the previous W. Germany), the population working in the environmental sector has increased by 26% in ten years (433 000 jobs in 1980 against 546 300 in 1990). The share in the total working population corresponding to these jobs increased, in the same period, from 1.7% to 1.9%. Predictions for 2000 estimate up to 748 000 the total number of environmental jobs (Kleinschmidt et al., 1994, p. 5). It is appropriate to add to these estimates, for the period 1990 – 2000, the

creation of 300 000 jobs in the Länder of the previous E. Germany (Kleinschmidt et al., 1994, p. 5).

Table 11.2 Direct and indirect environmental employment in France and in Germany.

	FRANCE (1994)		GERMANY (1990)	
		In %		In %
Direct jobs	275 000	*64.6%*	206 000	*38 %*
Indirect jobs	151 300	*35.4%*	340 300	*62 %*
Total *(In % of the total working population)*	426 300 *(1.8%)*		546 300 *(1.9 %)*	

Sources: Données économiques de l'environnement 1992-1993, 1993, p. 84.
Kleinschmidt & al, 1994, p. 3.

The distribution of direct environmental jobs following the domain of intervention is given, for France, in the following table:

Table 11.3: Distribution of direct jobs in the area of environment in France.

In thousands of persons.

	Direct jobs	
	1992	***1994***
Mobilization of water resource	36	35
Purification	57	59
Air	11	11
Noise	23	22
Collection and processing of wastes	82	85
Recovery	22	23
Ecological patrimony	15	15
Framework of life	21	22
Non distributed (administrative personnel)	3	3
TOTAL	**270**	**275**

Sources : Données économiques de l'environnement, 1992-1993, (1993), p. 84.
Données économiques de l'environnement, 1994-1995, (1996), p. 162.

The inclusion in the direct environmental jobs of 36000 units under the denomination 'mobilization of the water resource'[6] is doubtful because this activity cannot actually be considered either as a nature protection activity or as a processing pollution activity.

In Germany, the distribution of environmental jobs is not operated, as in France, following the domain of intervention but following a vertical decomposition of the economic activity close to the classical national accounts decomposition (Table 11.4). Due to this heterogeneity of data between the two countries, any direct comparison will be difficult.

Table 11.4 Distribution of environmental jobs in Germany.

Years *Sectors*	*1980*	*1984*	*1990*
Agriculture, forestry, fishing	1.9	1.9	1.5
Manufacturing and mining industry	156.8	156.9	196.8
Building	61	47.6	63.4
Services	124.9	127.7	171.3
Public utilities, administration (Local communities, States, Federal) and non-profit organizations	89.3	98.9	113.3
TOTAL	**433.9**	**433**	**546.3**

Source : Données économiques de l'environnement 1992-1993 (1993), p.86.

Let us note some points concerning the recent evolution of the environmental jobs in the two countries :

- The majority of environmental jobs are situated in the downstream side of the production process, that is they are associated with the processing of pollution. In France, approximately 60% of direct jobs (167 000 persons for the year 1994) are concentrated in the recovery and the processing of wastes (coming from manufacturing and households) as well as in the water purification activity. In the future, the evolution of the employment in these two sectors will depend on both the volume of wastes to process (see 11.3.3 below) and the gains of productivity that will take place in these activities. Currently, they hire a majority of unskilled employees.

[6] It is activity which the aim of capturing and stocking a natural resource : water

- These jobs are increasingly situated in the tertiary sector. One observes an important development of the activities of services linked to the environment: engineering, systems management, recovery, maintenance, etc.
- In France, local communities employ about 120 000 persons in environmental activities (direct jobs). At the end of 1993, 45 000 of these jobs (i.e. 37.5%) belonged to the category of C.E.S.[7] jobs (approximately 11% of the total C.E.S. jobs).

Public authorities also develop what is called 'green contract' jobs. Following the law of 27 July 1993, 300 million Francs have been allocated by the State in different projects which are job creators (of that amount 200 million has been allocated to Regions and 100 million to Regional Divisions of the Environment Ministry). From the beginning of 1994 up to January 1995, some 1 242 projects had been financed, representing globally 14 726 jobs under 'green contract'. Note, however, that among these jobs 75% were equally C.E.S. jobs.

This government policy of environmental jobs financed by public funds contributes to what is officially called *maintenance of nature,* but lies within the larger public measures framework adopted in France, known as *social treatment of unemployment.* Anyway, these jobs, belonging to the category of services, are situated essentially outside merchant activity.

The same tendency, although to a lesser extent, is observed in Germany, where programs of job creation have environmental protection as objective. To this end, in Länder of the previous E. Germany 117 000 environmental jobs have been created during 1991 and 1992 (Barde, Potier 1996, p. 20).

- The development of environmental jobs is also explained by the emergence of a new economic sector : the '**eco-industry**'. The statistical definition of this sector is again not fixed. In fact, neither the content nor the outline of this sector are subject to agreement on the international level.

The term eco-industry (or environmental industry) usually encompasses the pollution abatement activities intervening in the different large domains such as water, air, wastes, ground, noise, etc. However, the term eco-industry sometimes has a wider meaning:

> "eco-industries (...) can be described as the set of firms producing goods and services necessary to measure, prevent, limit or to correct damages caused to the environment, such that pollution of water, air, ground as well as problems linked to wastes. They

[7] C.E.S.: Job-Solidarity Contracts

> adopt clean technologies capable to reduce pollution and the use of raw materials to the minimum"
>
> (European Commission, 1994).

This sector employed in 1992, according to the estimations, 110 000 persons in France against 171 000 in Germany (Vickery and Iarrera 1996, p. 30).

The above figures on environmental jobs suggest that the overall environmental policy induces a net creation of jobs. In Germany, for the period 1980 - 1990, the loss of 55 000 jobs, due to intensified environmental regulation, is more than compensated by the creation of jobs through the development of the environmental sector (+113 000 jobs). For the nineties, the net gain is estimated at more than 185 000 jobs (Kleinschmidt et al., 1994, p.4).

In summary, the creation of jobs in the area of environmental activity takes place nowadays in two distinct forms:

- On the one hand, through public expenses programs resulting in the creation of subsidized jobs. The subsidies are granted by the State to private or public organisms which are thus engaged in an environmental protection activity. However, efforts are sometimes made in order to assure, progressively, the profitability of these jobs without public assistance. Thus, in Germany, governmental assistance is provided to sustain the transformation of the jobs originally created in the non-merchant sector into competitive jobs by means of firms creation aid. In this aim, an annual budget of 86 million DM has been allocated since 1993 (Barde, Potier (1996)).
- On the other hand, through the development of a merchant sector whose activity is the prevention and mainly the processing of pollution. No doubt its development is due to the implementation of pro-environmental measures.

However, the evolution of employment related to this last sector should be considered in relationship to the industrial orientations taken by the firms involved and to the specialization choices at the international level.

11.2.2. Eco-industry activities in France and Germany.

Currently, the environmental sector takes an increasing place in the economic activity of the industrialized countries. Even if definitions of this sector adopted by different national authorities are very heterogeneous (Vickery, Iarrera 1996, p. 45), different international organisms agree in announcing a strong growth for the environmental world market (5 to 6% per year up to the end of the nineties) as Table 11.5 shows.

Table 11.5 Growth perspectives for environmental market.

In billions of *$ USA*

	OECD[8] (1992)		**ECOTEC** (1994)			**ETDC**[9] (1994)		**Environmental Business International** (1994)	
	1990	*2000*	*1992*	*2000*	*2010*	*1990*	*2000*	*1992*	*1998*
World total	200	300	210	320	570	255	580	295	426
Europe	66	99	65	98	167	78	213	108	159
France	10	15	-	-	-	10	30	-	-
Germany	17	23	-	-	-	21	65	-	-

Source : OECD, (1996), "L'industrie mondiale des biens et services environnementaux", p.11.

From this table one can realize the importance of the European market within the world environmental market (between 30 and 40%). Also, one can notice the relative weight of the German market within the European one. It is important for the following arguments to show some closer differences of the activity related to the eco-industry between France and Germany.

11.2.2.1 French eco-industries. The French eco-industry market was evaluated about 10 billions $ USA in 1992 - 15% of the European market- (Vickery and Iarrera 1996, p. 11). In 1991, the turnover of environmental activities amounted to 30 billions of Francs (see Julia 1994 and Simond 1993). The following figure indicates the distribution of French eco-industrial activities turnover by domain of intervention for 1991.

[8] Clean technologies are not taken into account in this study. The same holds for the ECOTEC study. For ETDC, only substitution clean technologies is taken into account, with the exclusion of newly clean processes, production of clean energy and clean products. Environmental Business International study takes into account only some clean technologies.

[9] ETDC : Environmental Technologies Development Corporation.

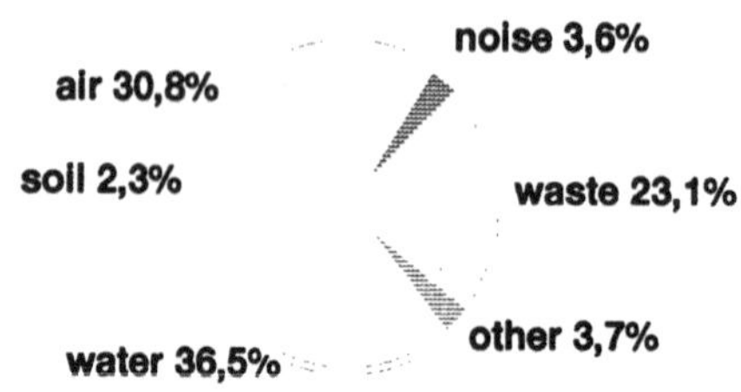

Figure 11.1 French ecoindustrial market (1991) (in % of ecoindustrial turnover)

Source : JULIA, M. (1994), *Les marchés de l'environnement*, SESSI.

Manufacturing industry and local community units constitute the two main customers of eco-industry. The demand of local communities is principally concentrated on equipment and services for water resource processing supplied only by some large firms. The manufacturing industry generally addresses its demand to the small and medium enterprises. In 1991, eco-industrial exports amounted to 9 billion Francs (30 % of their total turnover) including services (see also 11.3.2.2. below). The European market represents 69% of these exports (See Julia 1994, p. 42).

One of the central characteristics of the French eco-industrial firms is to propose goods and services directed to pollution abatement. In other terms, their intervention is rather specialized in the downstream side of the production process such as processing of water resources, sewage and waste management.

11.2.2.2 German eco-industries. The German eco-industrial market has developed rapidly during the eighties to attain currently a critical size on a quantitative and also qualitative plane inside the European and the world market (Kleinschmidt et al. 1994; Vickery and Iarrera 1996; Simond 1993). German eco-industries are specialized in advanced technology products intended either to measure and analyze pollution or to prevent and limit it at the source. One can thus identify a clear orientation in this country towards the design and the production of capital goods permitting a cleaner transformation process.

Germany is very competitive in almost all environmental activities. One denotes, nevertheless, a relative weakness in the absence of experienced private firms in operating services (water resource management, for example) with the exception of the recovery and the wastes processing services (Defeuilley and Quirion 1995).

The significant growth of the activity and employment linked to environmental protection allows us to state that, currently, the efforts deployed by the public authorities to resolve the ecological problems resulted in the emergence and the consolidation of an environmental industry which is, more and more, supported by a *solvable* demand inside or outside the economy. The existence and the increasing strengthening of this demand mean that jobs created in the corresponding merchant sector are potentially more viable than those created in the fundamentally subsidized non-merchant sector.

However, as we have just seen, there are some differences in the tendencies followed by the eco-industry in France relative to those followed in Germany. So, one can naturally raise the question of the incidence of these differences on the environmental employment in each of the two countries. This will be done in the following section.

11.3 Reformulating the jobs-environment relationship

In order to deepen the relationship between employment and environmental protection activities, it is helpful to introduce a new distinction inside these activities. This distinction, firstly, will allow us to propose a redefinition of what is called the 'central nucleus' of environmental activities. The expression 'central nucleus', according to the definition of the OECD, groups together the different economic activities that intervene downstream in the production process[10]. Secondly, this distinction can serve as a theoretical basis to predict the evolution of the structure of the above activities. This will be the object of the Section 11.3.1.

Thereafter, we propose some indicators that can differentiate the type of evolution of the environmental activities in France and in Germany. We claim that tendencies observed in the external trade of environmental goods and services are determinant for the evolution of the environmental jobs in the two countries (Section 11.3.2.). Finally, we show that the more the production (and international specialization) choices of a country are qualitative growth oriented the more important will be the gains for this country in terms of employment.

[10] Vickery and Iarrera (1996), p.8. "The nucleus of this industry is constituted by a group of goods and services meant to make less polluting the production and techniques process (equipment or downstream technologies), to treat water and sewage, and to fight against atmospheric and noise pollution".

11.3.1 The evolution of the central nucleus of environmental activities.

It is possible to decompose environmental activities in the following manner. On the one hand, we distinguish a first category of industries and services linked to the processing of the pollution (air, water, wastes, soil, etc.) that intervene downstream in the production process. This category groups the whole range of activities producing goods and services that go from the development of pollution processing technologies up to wastes recovery and their stockpiling in dumps. All of these activities can either exist inside polluting firms under vertical integration form or be undertaken by companies specialized in the pollution processing. In fact, we find here the bulk of what has been called eco-industry activities in Section 11.2.2. The OECD calls these activities the 'central nucleus' of the environmental sector.

They are justified by the restoration of damages caused to the environment by the overall production and consumption activities and, therefore, by the reply which the community gives to the production of harmful things in excessive quantities. But, even if they participate in the creation of added value, these activities are concerned only with goods whose production is not always compatible with qualitative growth. Even more, these goods constitute a sign of productive inefficiency (see 11.1.2).

In other words, the activities grouped under the name of 'central nucleus' should be reduced considerably if the economy turns towards a qualitative growth because, following the definition of such a growth, polluting firms are obliged, at least to some degree, to resolve pollution problems at the source.

On the other hand, a second category gathers pro-environmental activities that take place upstream in the production process. They provide the different economic sectors with cleaner production techniques, thrifty in natural resources and energy and causing lesser pollution. Therefore, they allow the improvement of firms' productive efficiency in the sense of M. Porter and C. Van Der Linde (1993, 1995). These activities can develop in two ways:

- by vertical integration: firms can decide to allocate R&D resources in the improvement of cleaner production techniques and the conception of new products implying less quantity of waste.
- by the emergence of specialized firms devoted to the development and the production of cleaner technologies.

F. Duchin et al. (1995) propose the term '***industrial ecology***'[11] to designate more specifically this type of activities. This term must be distinguished from that of eco-industry which is currently used in France

Note that 'industrial ecology' activities include equally all activities that participate to the ecological improvement of agricultural production processes, notably those that allow the avoidance of the ground depletion and the pollution of groundwater tables.

The preceding decomposition can be illustrated in Figure 1. The upper part of the diagram represents the activities intervening upstream in the production process and the lower part those intervening downstream. Inside each one of these two categories, we further separate the activities according to whether they are oriented towards the creation of new techniques (innovation) or the effective use of such techniques (application).

The industrial ecology develops more or less rapidly following the demand shifts for cleaner production techniques and the evolution of pollution abatement activities. Correspondingly, the activities of firms specialized in the development and the production of end of pipe processes are reduced progressively. However, these activities are not totally compressible. In fact, some of them will always exist as, for example, the production of water purification processes.

Let us stress the following point. For a given degree of public environmental regulation, the growth of the industrial ecology activities depends, of course, on the level and the quality of the economy's innovation process and capacities. However, in an open economy, industrial ecology innovations can be imported and applied directly by firms. Thus, qualitative growth is rendered possible through innovations stemming from two competing sources. In other terms the development of a significant national industrial ecology is not a necessary condition for such a growth.

Now, the development of activities intervening upstream in the production process will take place at the expense of activities belonging to the downstream category of activities. Thus, the central nucleus contains activities that, while contributing to the environmental protection, take place in the upstream phase of

---

[11]Duchin et al. (1995). The authors define the term of industrial ecology as the industry which "prevents environmental degradation instead of capturing pollutants through end-of-pipe solutions after they are generated. [...] One can distinguish two different branches of Industrial ecology: Eco-restructuring is associated with the conviction that new technologies will make possible the dematerialization of production to only several percent of current material requirements, and the decarbonization of the energy system, most likely through the transition through natural gas on hydrogen. Design for environment is the branch of industrial ecology associated with the design and implementation of solutions that are pratical within a 5 - 10-year period".

the production process. This evolution amounts to a *structural change* of the central nucleus.

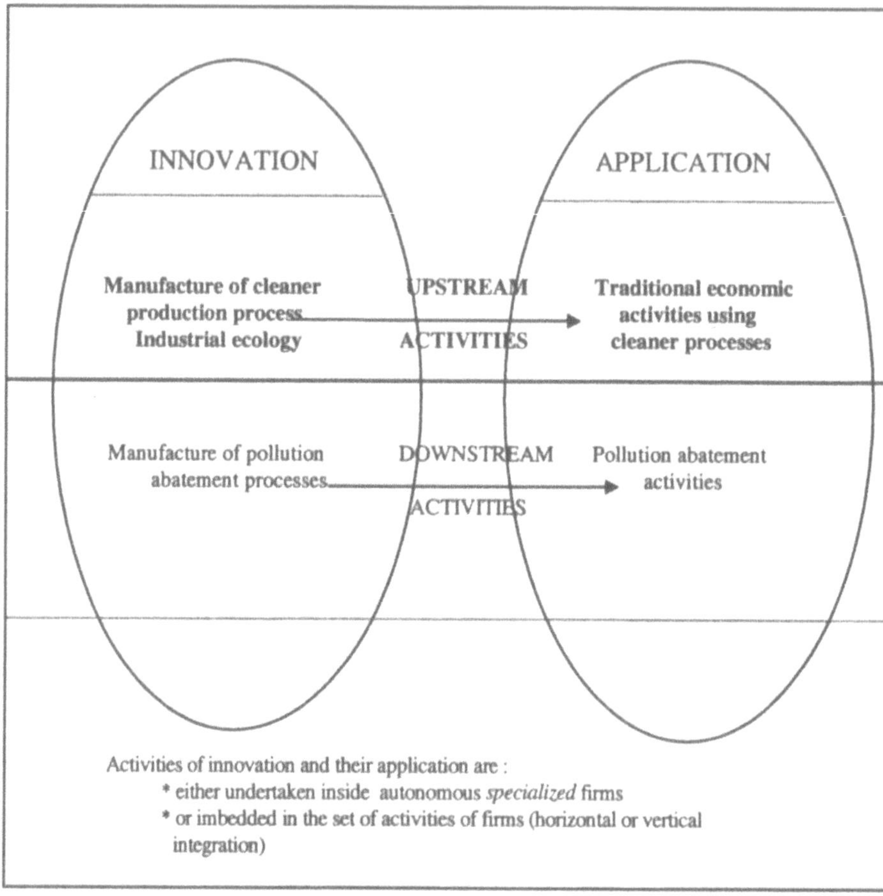

Figure 11.1 Aggregate categories of environmental activities in the economy.

To clarify this structural change, we construct a second diagram, inspired by the 'input-output' tables method of analysis. It allows us to point out the productive connections between the different economic sectors using environmental goods and services. The set of these goods and services includes the innovative production techniques.

The economic activities appearing in column are divided in those oriented towards the elimination of polluting wastes at the source and, generally, the increase of the productive efficiency of resources and those that favor pollution abatement. The rows classify the different goods and services following their destination to the upstream or downstream environmental activities.

Figure 11.2 Utilization of environmental goods and services by the different economic activities

	Activities / *Goods and services*	Activities reducing pollution to the source	Activities of abatement: Urgently interven-tion	Processing of the water	Sewage's and reorientation wastes	Air pollution	Noise pollution	Soil pollution
Goods and services type industrial ecology	Cleaner production equipment	x	0	0	0	0	0	0
	Equipment improving the energy-efficient	x	0	0	0	0	0	0
	R-D	x	0	0	0	0	0	0
	Training and education to the environment	x	0	0	0	0	0	0
	Services	x	0	0	0	0	0	0
Goods and services participatin g to the pollution processing	Pollution abatement equipment	0	x	x	x	x	x	x
	R-D	0	x	x	x	x	x	x
	Services, maintenance unit	0	x	x	x	x	x	x
	Training and education to the environment	0	x	x	x	x	x	x

Legend : x : positive flow in goods and services (lines) in direction of activities (columns)

0 : zero flow

The potential structural change of environmental activities will take place if economy turns towards a qualitative growth. **It is at the same time a qualitative and quantitative change.**

- Qualitative, because globally, the nature of environmental activities is modified. The 'central nucleus' of these activities is going to contain increasingly activities of industrial ecology (see figure 11.1).
- Quantitative, because activities situated to the north-west of the table are going increasingly to use goods and services of industrial ecology. In fact, that implies a relative decrease of pollution abatement activities situated to the southeast of the table. However, this decrease is not inevitably compensated, for a given country, by a increase of environmental innovation activities upstream of the process of production. Indeed, importing cleaner

processes, the economy can equally turn towards a qualitative growth (see 11.3.2.2 below).

This growth will take, in general, different forms and intensities per country. Innovation paths and diffusion processes will also be different. Moreover, insofar as this growth will not start at the same time, dynamic consequences will appear: countries that first developed an appropriate innovation system will possess competitive and more general comparative advantages than the countries that were left behind (see. 11.3.2.2 below).

11.3.2 Environmental activities in an open economy

Statistical observation allows us to illustrate a significant difference between France and Germany, in the orientation that their environmental activities have taken in the last decades. We use three indicators[12] : the importance of the pollution abatement investment, the public expenses in R&D and the external trade of environmental goods and services.

11.3.2.1 Differences in the pollution abatement investment and in the R&D activity.

A first indicator of the difference in the orientation towards qualitative growth between France and Germany is given by the evolution of the pollution abatement investment.

In order to use this indicator it is necessary to distinguish between investments in cleaner production processes and those intended to the pollution processing that is the end of pipe. Statistics convey three categories of pollution abatement investments (see Quivaux, Sabot, Vimond 1995) :

- ***Investments linked to risk prevention.*** These are destined to prevent or to limit accidental pollution.
- ***Specific investments.*** These aim to reduce pollution linked to the usual production process. It is related with, for example, filters, dust controls systems, purification stations, etc.

---

12 This choice is relatively arbitrary. Despite of the lack of available statistics capable of showing a difference in the orientation of environmental activities between France and Germany, other indicators could be chosen, as for example, statistics on international trades of industrial wastes. See Scheers (1995) for a study of these trades between France and Germany.

- ***Investments allowing improvement in production techniques*** (the corresponding statistics display the environmental share of the total investment dedicated to the improvement of the production techniques). These entail, in opposition to the second category, a renewal of the production tool which becomes more respectful of the environment.

Figures 11.4 and 11.5 illustrate the proportions of these three investment categories for France and Germany, respectively (unfortunately, we have no data for 1991 for Germany).

In France, these investments represented, in 1993, 5.8 billion Francs, (3.3% of the total investments of industry). As compared to those of 1992, their increase was 5 or 6%. But, in 1991, they represented 5.9 billion Francs. The financial effort linked to the improvement of processes has decreased by 1 billion between 1991 and 1993. On the contrary, specific investments have increased by 0.9 billion Francs during the same period. Note that, inside this specific investment category, the share of investments going to the water and sewage processing activity has been 52% in 1991 and only slightly less (49.6%) in 1993.

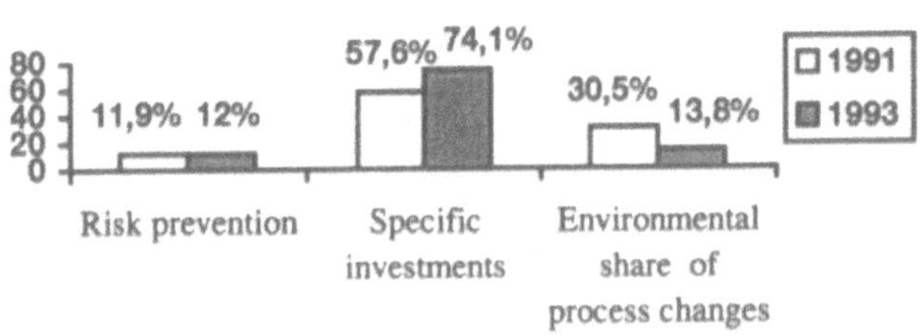

Figure 11.4 Pollution abatement investment in French industry (1991 and 1993).

Source : QUIVAUX,. SABOT, VIMOND J (1995), *Les investissements antipollution, édition 1995*, SESSI.

In Germany, the share of pollution abatement investments represented, in 1993, 5.6%[13] of the total industrial investments (50% more than the effort consented by industrial French firms, see Quivaux, Sabot and Vimond 1995, p.7).

Thus, the environmental share of process improvement in the total abatement pollution investments is appreciably more important in Germany than in France. This share represented, in 1993, 29.8% in Germany against 13.8% in France.

A second indicator of the qualitative growth orientation is the effort made by the two countries in favor of R&D[14] devoted to environmental protection. Public budgetary credits in R&D allocated to environmental programs were, in 1992, 150.4 million $ US in France against 529.2 million $ US in Germany. These amounts corresponded respectively to 1.1% and 3.6% of the total public credits allocated to the R&D (Vickery and Iarrera 1996, p.39).

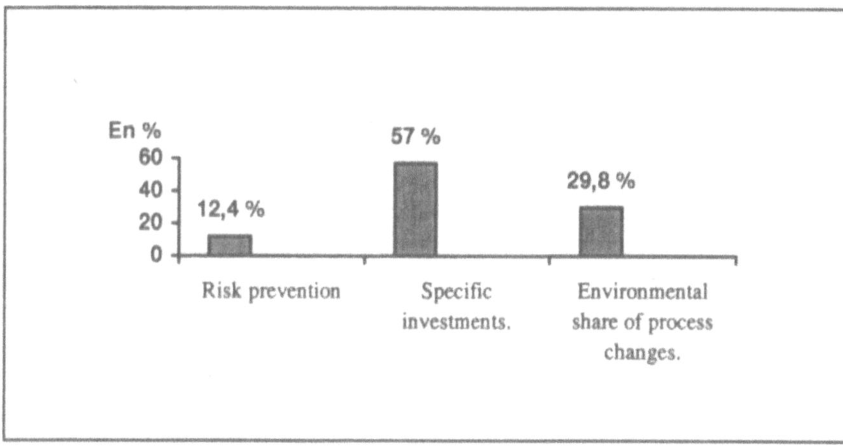

Figure 11.5 Pollution abatement investment in German industry (1993)

Sources : Quivaux, Sabot, Vimond1994. Kleinschmidt & al. (1994).

Even if it is actually impossible to know precisely the share of public R&D funds devoted to the upstream solutions and that directed downstream, one can find some structural differences in the allocation of these expenses between the

[13] Pollution abatement investments in the Länder of ex E. Germany represented, in 1993, 9.6% of the total industrial investments. See Quivaux, Sabot, Vimond 1995, p. 7.

[14] The analysis of the environmental R&D effort is difficult because of a lack of available statistics about this subject.

two countries. For example, waste processing is one of the four large sectors where France has concentrated its operational R&D programs. On the other hand, in Germany, the development of environmental technologies allowing cleaner production is strongly supported by public authorities (see OECD, 1992).

11.3.2.2 Differences in the international specialization. The environmental industry is becoming increasingly internationalized. Currently, world exports of this sector represent a percentage comparable to those of the other traditional manufacture sectors. In terms of absolute magnitude, they are equivalent to those of the pharmaceutical industry (Vickery and Iarrera 1996, p.26).

Germany is among the first world exporters of environmental goods and services. Its balance of trade displays a surplus of 720 million $ US. The corresponding amount for France is a surplus of 14 million $ US.

Moreover, one can assert some neat differences in the specialization of the two countries. France is, clearly oriented towards external pollution abatement markets. And more precisely, this orientation is illustrated by international successes of its firms in the water and sewage processing market. Another market also invested in by French firms is that of nuclear and industrial wastes (Scheers 1995). Consequently, France specializes in downstream environmental activities rather than in the upstream ones.

Germany is apparently oriented towards the improvement of existing production processes and especially towards the development of cleaner new production techniques. This orientation can be attributed to at least two reasons:

- first, this country has had an technological advance, for a long time, in the production of capital goods (machines-tools, intermediate goods, etc). Correspondingly, this know-how has benefited Germany when the environmental protection norms came into force.
- second, the important needs in pollution abatement due to a highly spatially concentrated industrialization.

One can assert that such a development of the German environmental sector is the result of an interaction between producers and users bringing about intense innovation processes which is a source of competitiveness in the long term (Fagerberg 1995 ; see the thesis of S.B. Linder 1961). The fact that German firms produce for the national market results in gains in terms of apprenticeships and economies of scale which in turn translates into a strong international competitiveness of this sector.

Thus by intensifying its production of capital goods, not only allowing pollution abatement but also cleaner production, Germany turns towards a qualitative growth. However, in a dynamic perspective, the possibility for German firms to exploit foreign markets gives them additional financial resources which in turn consolidates the orientation of the German economy in its qualitative growth path. In sum, aside from the initial role of society's demand for such a growth, successful external trade has certainly a role in reinforcing the development of activities of industrial ecology.

11.3.3 Consequences on the environmental jobs : prospective propositions

The possible orientation of economies towards qualitative growth, in accordance with the sustainable development idea, raises the question of employment with a new acuteness. Indeed, as we showed in 11.3.1, such an orientation is characterized by a displacement of the 'central nucleus' of the environmental activities, which means a decline in jobs linked to the processing of pollution. This does not go therefore in direction of an improvement in the unemployment situation, as some authors have sometimes suggested (see, for example Oates 1995).

Even more, such a result strengthens, in a different way, the opposite idea, more commonly admitted in the economic literature, advocating the incompatibility between environmental protection measures and employment. In fact, we show that during the qualitative growth process, the jobs-environment relationship is more complicated than it appears at first sight.

11.3.3.1 Qualitative growth and creation of jobs. In the short and medium term, it is possible to support the idea of an important growth of the category of environmental jobs that are linked to pollution abatement. Indeed, in order to respond rapidly to the social demand for an improvement in the life framework, public authorities are apt to favor, in the short term, the development of pollution abatement and the maintenance of nature activities. Thus, because of the usual irreversibility of public budget expenses, as well as phenomena related to firm's sunk costs, the environmental activities are susceptible to a trajectory of downstream activities. In consequence, the jobs created in the environmental sector belong in the same downstream category.

However, if a clear objective of qualitative growth is adopted in the society and above all is favoured by the authorities, the evolution of the environmental jobs will take another trajectory. This evolution must be appreciated in a long term perspective. More precisely, it is necessary to consider these jobs in relation with the speed of displacement of the 'central nucleus' of the environmental activities.

- On the one hand, the qualitative growth is based on an increase in the productivity of resources and therefore on an improvement of the firms' international and national competitiveness. This improvement entails an increase in national wealth and can thus translate into a net creation of jobs. Nevertheless, these jobs are not situated in the environmental sphere in the sense in which it is conceived generally. In fact, these jobs are spread throughout the different branches or sectors of the economic activity·
- On the other hand, the evolution of jobs in the environmental sphere will take different forms depending on the downstream or the upstream type of activities. The diminution of jobs, that will, surely, occur downstream, will be partly compensated by the increase of activities and therefore of related jobs that will take place upstream. At the same time, a qualitative change in the nature of jobs is due: skilled jobs are going to substitute for jobs of inferior qualification level.

We underline, however, the role that the international specialization of the economies can play in the evolution of employment (see 11.3.2.2). Firms of countries which are first oriented towards qualitative growth will certainly gain market share by providing cleaner production processes to firms of countries which are 'less advanced' in the direction of the qualitative growth. Thus, even if an open economy adopts a qualitative growth path, the fact of importing cleaner production techniques from abroad can prevent the creation of corresponding jobs. In that sense, the above mentioned substitution between the two categories of environmental jobs in a particular country is closely related with phenomena of international trade.

These considerations, emanating from our theoretical construction, will now be used in order to deepen the relationships between environmental policy measures and employment.

11.3.3.2 The coherence between the pro-environmental economic policy and the creation of environmental jobs. Faced with to the objective of job creation in the environmental sector, environmental economic policy can adopt one of the two following polar possibilities:

- The first consists, for public authorities, in encouraging firms to develop principally pollution abatement activities which, correspondingly, favours the creation of jobs. Yet, as we have already observed, this policy is not compatible with qualitative growth. In these circumstances, created jobs are not viable in the long run, unless the economy is specialized in the processing of imported wastes. But, even then, if, after some time, waste exporting countries themselves adopt a qualitative growth path, the related environmental jobs in the importing country are not viable.

- The second possibility is to build a strategy of transition based on the displacement of the 'central nucleus' of environmental activities according to the objective of qualitative growth. The duration in time of this transition depends on several factors; notably, intrinsic possibilities of growth of the economy as well as the intertemporal preferences of the society for environmental protection.

Note that, nowadays, a country's engagement in environmental protection, apart from institutional decisions emanating from internal factors, depends also, to a large extent, on international agreements. More particularly, the evolution of the environmental sector, for countries like France and Germany, depends largely upon the European harmonization process of environmental norms and regulations

What are the modes of this transition? Of course, in the long term, jobs linked to the downstream type environmental protection will decrease. This tendency will nevertheless be compensated by an increase in the work force for augmenting industrial ecology activities. At the same time, adaptation costs can be necessary, for example, in labor training and firm redeployment.

The transition strategy has to take into account the international trade of goods and services produced by the industrial ecology, beside that of wastes. Moreover, firms employed in industrial ecology activities can export their products. That allows them to develop their scale of production and to strengthen their economic profitability. Financial gains, realized to the level of firms, strengthen the orientation of the economy to the qualitative growth. In this field, the different national environmental economic policies should not neglect the benefits that can accrue globally by exploiting the complementarities and the differences existing between the countries.

We insist, in addition, on the fact that the development of industrial ecology activities consists in integrating environmental activities inside the merchant sphere[15], while, often, a non-negligible share of environmental activities takes place in the non-merchant sphere (subsidized activities and non-profit associations). This shift amounts to an increase in the financial autonomy of the

[15] Public subsides will certainly continue to exist, in particular for R&D.

whole of the environmental protection activities and correspondingly diminish the public budget related to this protection and the associated jobs.

In our view the most central point of this issue has to do with the allocation of financial resources (public or private ones) among the different kinds of environmental activities. The choice is not only between resources devoted to the industrial ecology activities and pollution abatement activities but also between the operating costs and the R&D expenses. A country which neglects the financing of upstream environmental production processes induces national firms, wishing to improve their productive efficiency (see 11.1.2), to provide cleaner processes by way of imports. However, this competitive behaviour of private firms can lead the economy towards some degree of qualitative growth and the consequences of such behaviour result in a decrease of environmental jobs in the downstream activities.

Thus, if a country is hindered in the domain of industrial ecology activities, relative to other countries, this country can, in a more or less long time period, find itself in an '*unemployment trap*': *the diminution of jobs downstream cannot be compensated by an increase of jobs upstream.*

Conclusion

Economic activities linked to environmental protection keep a relatively small share of the working population occupied. But, in a general context of increasing unemployment, the absolute and relative number of these jobs increases in most western industrialized countries. This is the case in France and Germany.

The evolution of all these jobs does not follow the same logic. Thanks to the qualitative growth concept we have been able to operate a distinction in kind between environmental activities and the associated jobs.

On the one hand, a part of these activities intervene upstream in the production process, thus participating in the elimination of pollution at the source. The development of industrial ecology activities then has to bring an increase of jobs in the economy.

On the other hand, a certain number of activities intervene downstream in the production process. The orientation of the economy towards a qualitative growth implies a diminution of these activities and therefore of the related jobs.

The observation of the evolution of these two categories of activities and associated jobs is dependent on the statistical system that countries have or are going to adopt in the future. Unfortunately, available statistics are not sufficiently detailed to allow a precise differentiation of these activities as well as of the corresponding jobs. Such a lack of statistical information reveals the currently preeminent conception of environmental activities traditionally concentrated at the downstream side.

The orientation of the economies towards qualitative growth renders inescapable the displacement of the center of gravity of activities from the downstream side to the upstream one. The preceding analysis suggests that it is necessary to anticipate and to control the evolution of the nature of these activities and the associated jobs. The same analysis can give some indicators as to the improvement of the statistical system.

We have shown that the jobs-environment relationship is less straightforward than it is commonly admitted. Public authorities should further rationalize the allocation of resources between the encouragement of pollution abatement activities and innovative activities dedicated to environmental protection. It is also suggested that the commitment of resources and job creation in downstream activities for income redistribution reasons not only can be inconsistent with qualitative growth, but, moreover, it may turn out to be against the unemployment reduction objective.

References

Audinet P, Conséquences sur l'emploi des choix liés aux décisions en matière d'environnement. Etude bibliographique et synthèse thématique, Document de travail, C.I.R.E.D, 1993.

Barbera A, McConnell V, 'The impact of environmental regulations on industry productivity : direct and indirect effects', Journal of Environmental Economics and Management, 1990; 1 : 50 - 65.

Barde JP, Potier M, 'Un effet vert sur l'emploi', L'observateur de l'O.C.D.E.,1996; 198 :17 - 21.

Benhaïm J, Schembri P, 'Environnement et orientations stratégiques du changement technique', Economie Appliquée, 1995; 3 : 39-69.

Berkes F, Folke C, 'A systems perspective on the inter-relations between Natural, Human-made and Cultural Capital', Ecological Economics, 1992; 5 : 1-8.

Blazejczak J, 'Environmental policies and foreign investment : the case of Germany', in Environmental Policies and Industrial Competitiviness, O.C.D.E., Paris, 1993.

Comolet A, Weber JL, 'Un instrument de connaissance et d'aide à la décision : le système de comptes de patrimoine naturel français', Revue Economique, 1990; 2 : 243 - 267.

Commission Européenne, 'Industries écologique dans l'UE' in Panorama de l'industrie de l'UE 94, Bruxelles, 1994.

Commission Mondiale sur l'Environnement et le Développement (CMED), dite commission Bruntland, Notre avenir à tous, Ed.7 du Fleuve, Montréal, 1988.

Daly H, Steady state economics, Second Edition, Island Press, Washington DC, 1991.

Defeuilly C, Quirion P, 'Les déchets d'emballage ménagers : une analyse économique des politiques allemande et française', Economie et Statistique, 1995; 290 : 69-79.

Dobre M, Structure et profils des emplois environnement en France, Institut Français de l'Environnement, Travaux et Recherches, 1993, 1.
Dobre M, Hammer B, 'Opinion publique et environnement : une préoccupation toujours forte mais pas à n'importe quel prix', Les Données de l'Environnement, Institut Français de l'Environnement, 1995; 14.
Duchin F, Lange G-M, Kell G, 'Technological change, trade and the environment', Ecological Economics, 1995; 2 : 185 - 193.
Fagerberg J., 'User-producer interaction, learning and comparative advantage', Cambridge Journal of Economics, 1995; 19 : 243 - 256.
Faucheux S, Froger G, Noël J-F. 'Quelle hypothèse de rationalité pour le développement soutenable' Economie Appliquée, 1993; 4 : 59 - 103.
Georgescu-Roegen N, La Décroissance ; Entropie-Ecologie-Economie, Editions Sang de la Terre, Paris, 1995.
Hartwick JM, 'Intergenerational equity and the investing of rents from exhaustible resources', American Economic Review, 1977; 5 : 972 - 974.
Heraud J-A, Llerena D, 'Environnement et traditions nationales : comparaison et interprétation socio-économique des politiques publiques et des stratégies industrielles en Europe du Nord', Economie Appliquée, 1992; 4: 45 - 75.
Jacobs M, Green jobs? The employment implications of environmental policy, a report to WWF, CAG Consultants, Brussels, 1994.
Jaffe A, Peterson S, Portney P, Stavins R, 'Environmental regulation and the competitiveness of U.S. manufacturing : what does the evidence tell us ?', Journal of Economic Literature, 1995; 1 : 132 - 163.
Julia M, Les marchés de l'environnement, Ministère de l'Industrie, des Postes et Télécommunications et du Commerce Extérieur, Direction générale des stratégies industrielles, Collection SESSI études, 1994.
Kleinschmidt M, Nordhause J, Rehfeld D, Relationships between environmental protection and employment in the chemical industry. A brief survey on the current state of the debate, Working paper Gelsenkirchen, 1994.
Linder SB, An Essay on Trade and Transformations, Uppsala, Almquist & Wicksell, 1961.
Ministère de l'Environnement, Données economiques de l'environnement 1994/1995, Edition Economica, 1996.
Ministère de l'Environnement, Données Economiques sur l'Environnement 1992-1993, 1993.
Medhurst J, 'Environmental costs and industry competitiviness', in Environmental Policies and Industrial Competitiviness, OCDE, Paris, 1993.
Oates A, Overview of the European debate on environment and employment, Working paper for SAMF-Tatung -Ökologie und Beschäftigung-, Gelsenkirchen, 1995.
OECD, Examen des performances environnementales : Allemagne, Paris, 1993.
OECD, Environmental policies and industrial competitiviness, Paris, 1993.
OECD, 'Technologie et environnement', in La technologie et l'économie : les relations déterminantes, OECD, Paris, 1992.
Palmer K, Oates WE, Portney P, 'Tightening environmental standards : the benefit-cost or the no-cost paradigm ?', Journal of Economic Perspectives, 1995; 4 : 119 - 132.

Passet R, 'L'aménagement du territoire dans une perspective de développement durable', Revue d'Economie Régionale et Urbaine, 1994; 4 : 523 - 535.

Pearce D., 'Les entreprises et les valeurs de l'environnement', in Environnement et choix économiques d'entreprise, Stoffaës C., Richard J.-M. (eds), Cahiers de Prospective, InterEditions, 1995.

Pezzey J, Sustainable development concepts ; an economic analysis, World Bank Environmental Paper, n°2, 1992.

Porter ME, Van Der Linde C, 'Towards a new conception of the environment-competitiviness relationship', Journal of Economic Perspectives, 1995; 4: 97 - 118.

Quivaux R, Sabot P, Vimond J. 'Les investissements de l'industrie protégeant l'environnement affectés par la récession en France comme en Allemagne', Le 4 Pages, SESSI, 1994; 37.

Quivaux R, Sabot P, Vimond J, Les investissements antipollution, édition 1995, Collection référence les chiffres clés, SESSI, 1995.

Renn O, A regional concept of qualitative growth and sustainability. A pilot project for the German State of Baden-Wüttemberg, Arbeitsbericht Discussion Paper, Akademie für Technikfolgenabschätzung in Baden-Wüttemberg, n° 2, 1995.

Roch J. Environnement et emploi. Vers une observation statistique des emplois dans le domaine de l'environnement, Institut Français de l'Environnement, Notes de méthode, n° 5, 1996.

Scheers M-J. Une analyse économique des mouvements transfrontaliers de déchets industriels. Le cas de la France et de l'Allemagne. Quelles voies vers une harmonisation des politiques nationales?, Mémoire de DEA d'Economie Industrielle, Université des Sciences Sociales-Toulouse 1, 1995.

Simond A. L'industrie française de l'environnement : caractéristiques et marchés, Chambre de Commerce et d'Industrie de Paris, Direction des études, 1993.

Solow R. 'Intergenerational equity and exhaustible resource', The Review of Economic Studies, Symposium on Exhaustible Resources, 1974; 29 - 45.

Thomas J, Lavoux T, 'Comparaison France-Allemagne. Réduction du CO2 dans les années 80 : l'efficacité énergétique à la clé', Les Données de l'Environnement, Institut Français de l'Environnement, 1995; 15.

Tobey JA, 'The effects of domestic environmental policies on patterns of world trade: an empirical test', Kyklos, 1990; 2 : 191-209.

Tobey JA, Smets H, 'The polluter-pays principle in the context of agriculture and the environment', The World Economy, 1996; 1: 63 - 88.

Van der Linde C. 'Micro-economic implications of environmental regulations: a preliminary framework', in Environmental Policies and Industrial Competitiviness, OECD, Paris, 1993.

Vickery G., Iarrera M., L'industrie mondiale des biens et services environnementaux, OECD, Paris, 1996.

12 FROM THE EXPECTED TO THE DESIRED FUTURE OF PASSENGER TRANSPORT
A Stakeholder Approach

Sytze A. Rienstra and Peter Nijkamp

Abstract

Sustainability as an unambiguous policy goal is not a priori secured, as is clearly shown in the transport sector, where the negative externalities are still increasing despite official policies aiming at a reduction of these external costs and at the achievement of a sustainable transport system. To analyse why this is the case, a conceptual model is developed in this paper, in which stakeholders are identified which influence sustainable transport policies. These stakeholders are individuals, the public sector (subdivided into politicians and civil servants), international organizations and pressure groups (car industry, oil industry, car users and environmental groups). It appears that – although it may be assumed that nobody desires an unsustainable future – most incentives and mechanisms in our conceptual model of the decision-making process hamper the achievement of a sustainable transport system. In the second part – by way of empirical test – results of a questionnaire among Dutch transportation experts on the expected and desired future of European passenger transport are concisely discussed, in which the year 2030 is taken as a reference year. It appears that in the expected future the stakeholders largely behave as predicted in the conceptual model. In the desired future, however, the stakeholders have to change their attitudes drastically. It is concluded that a strategy focused on behavioural changes of stakeholders is necessary for the achievement of sustainability goals. Sufficient and appealing information, the existence of alternatives for car use and a cooperation of all stakeholders are key success factors for the achievement of a sustainable transport system in the future.

12.1 Introduction

In a simple atomistic society driven by market forces, individual citizens operate on the basis of optimizing their utility in a Pareto-optimal way. When no transaction costs occur, even externalities will be included in the market decision (Coase 1988; Nijkamp and Rienstra 1997). Of course, transaction costs do exist in the real world, so that in a somewhat more complex and ordered society a government may have to intervene in order to provide services and policies to cope with market failures, so that a more or less social optimum outcome is secured.

However, in practice, government policies and outcomes of decision-making processes in society are not that simple. Other stakeholders also play an important role, while each stakeholder has his own utility function. Also, individuals may not always seem to act in a rational way, while governments may have their own utility functions, which differ from the societal ones. Therefore, the outcomes of a certain policy-making process may not result in socially optimal outcomes. In this way, government failures occur, resulting in a suboptimal allocation of resources. When governments want to introduce regulative policy measures, a trade-off should be made between the benefits of solving market failures and the costs of government failures (Nijkamp and Rienstra 1997).

When the results of the very many attempts and policies to achieve more sustainability in transport are analysed, this observation seems to hold very well. Officially, policies aim at achieving sustainable transport, by reducing congestion and stimulating public transport. In reality, however, these policies are not very successful. In every European country the mobility levels increase, the modal split changes in favour of the car and airplane, and the emissions of harmful gases increase. Calculations indicate that the external costs of car use may be as large as 3–5% of the European GNP (Verhoef 1996), although it should be acknowledged that such calculations are only valid under certain assumptions. In fact, this is a strange observation, since almost no individual in society wishes the environment to detoriate. At the same time congestion – especially in large cities and densely populated regions – becomes an increasing problem, which is not coped with in an effective way.

Therefore, it seems that new policy solutions and strategies are hardly found for the transport system, so that the problems tend to increase. Apparently, there are forces which push the system in a direction which is not desired. In this paper we analyse this is the case and which changes are necessary in the decision-making process to achieve a more sustainable transport system.

The content of this paper is as follows. Section 12.2 a conceptual model of the decision-making process in the transport sector presents, which explains the question why this is the case. In Section 12.3 this model is tested in an empirical way by analysing the opinions of Dutch transportation experts on the expected and desired future of the transport sector. Strategic policy conclusions are drawn in Section 12.4.

12.2 A Conceptual Model of Stakeholders

In light of the above observations, it makes sense to identify stakeholders in the process for achieving sustainable transport, to analyse the incentives of these stakeholders and to explain why the transport system still moves away from sustainability. Such an analysis may be carried out by applying public choice theories to the transport sector, i.e. by applying an economic analysis to the outcomes of the decision-making processes which should – but do not – lead to a sustainable transport system.

There are two modes of policy analysis in public choice theory (Rietveld 1997). The normative approach analyses the optimal outcome of a process, by determining which outcome is Pareto-optimal. The second analysis is the positive approach, in which the actual outcome of the decision-making process is analysed. Here, we will focus on the second type of analysis, applied to sustainable transport policies.

In public choice theories, the basic assumption is that every human being acts as if he were a rational utility maximiser (see e.g., Dunleavy 1991). This assumption provides a clear basis for analysing the behaviour of individuals, although one also should pay attention to theories, which explain non-rational behaviour. There are, however, also attempts to rationalize this behaviour (see e.g., Brady et al. 1995).

Eventually, it may be possible to change incentives of the stakeholders in one way or another, so that current trends may bend into the direction, which is desired by society at large. A conceptual model which presents the influence of the main stakeholders on (sustainable) transport choices is presented in Figure 12.1.

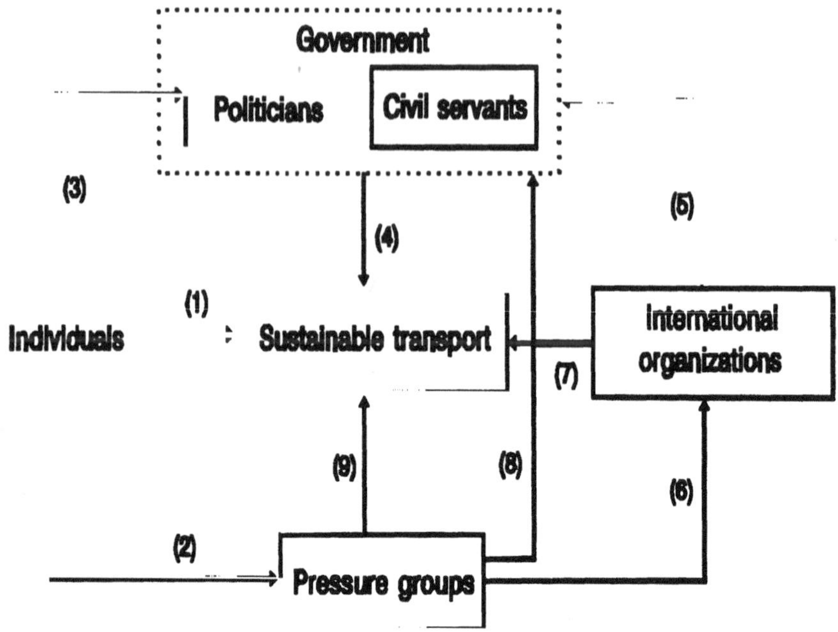

Figure 12.1 A conceptual model of stakeholders in a decision-making process towards sustainable transport

Only the strongest relations are presented in Figure 12.1; it should be acknowledged that of course other relations also hold, but these are assumed to be less strong, however. Policies to change e.g. the behaviour of individuals or the R & D of the car industry (which may be an important pressure group) are included in the 'sustainable transport' box. Next, we will discuss the behaviour, incentives and relations of the distinct stakeholders.

12.2.1 Individuals and sustainable transport

The first group of stakeholders in the transport system is formed by the individuals. In the end these are the consumers of transport; the behaviour of this group is therefore critical for the achievement of sustainability. This group of stakeholders may influence the road to sustainable transport directly, but also via other stakeholders as was already shown in Figure 12.1.

In the past decades, individualism has become a megatrend in society. Individuals seem to act more and more like 'calculating citizens', who are oriented towards monetary gains and their own societal position. The influence of e.g., churches, labour unions and other societal organizations has diminished. As a result, it seems that altruistic behaviour has decreased too, which is also reflected in public policies which tend to abolish e.g., social security systems, subsidies for social support, and subsidies for housing of lower income groups. A consequence is that it becomes more difficult for governments to steer the individuals, while (financial) incentives and market-based interventions are increasingly becoming more important. Individuals behave thus more and more as if they were rational (calculating) citizens, as is largely assumed in the public choice literature. This however, may lead to outcomes which do not seem rational, e.g., because acquiring information is not free (time and monetary costs). In this respect it should be added that information has become much cheaper in the past decades (radio, television), a trend which is still continuing (e.g., the development of 'the electronic highway'); therefore, individuals are becoming better informed. At the same time, the overload of information leads to the observation that the presentation of information becomes increasingly important; otherwise the information is lost because it is not appealing.

In this respect it should be acknowledged that even full information may not result in a societally optimal outcome, as is shown e.g. by the prisoners dilemma and other game theoretical examples like free rider behaviour (Frey 1983). Further, the utility of future generations may not be taken into account in the decision making process, leading to suboptimal outcomes from a long term economic and sustainability perspective.

In the decision-making process of individual stakeholders, information is the starting point. The decision is first of all determined by (perceived) alternatives. The final choice between these alternatives is made on the basis of a cognitive and motivational decision-making process, making use of the available knowledge of the alternatives, perceived or presupposed pros and cons, personal preferences and attitudes, and social and individual norms. In other words, the alternatives are traded off against each other, which is, however, only partly rational and very strongly biased (Rooijers and Steg 1991). It should, however, be acknowledged, that individuals in practice often have little choice in transport, because there are few alternatives. E.g., without any cycling infrastructure in cities, it is very hard and dangerous to cycle instead of drive a car, so that it is not a real option. In these cases there is a clear role for the government to provide such alternatives. The above-mentioned factors will now be concisely discussed, partly using the behavioural theories of Fishbein and Ajzen (1975) and Ronis et al. (1989).

Knowledge. An optimal rational choice between alternatives first of all includes that all possibilities are known in an objective way. This is often not the case, however. In transport, for example, those people who travel little or not at all by public transport hardly see it as an alternative. The knowledge from which a choice has to be made is usually just a subset of all feasible possibilities; this leads to the first subjective bias in the decision-making process.

Perceived pros and cons. An optimal objective choice further presumes that all pros and cons of the alternatives are known, which then have to be weighed on a rational basis. In reality, these conditions are not fulfilled, because they are both largely biased as a result of selective perception and cognitive dissonance mechanisms (Festinger 1954). People try to reduce this dissonance psychologically to the maximum extent by (Brady et al. 1995):

- revoking the decision when new information comes available;
- making their own decision more attractive;
- reducing the attractiveness of alternatives;
- avoiding obtaining new information;
- providing funds to e.g., an environmental organization which claims to have the solution to a problem.

As a result, people in general are inclined to overestimate the advantages and to underestimate the disadvantages of their own way of behaviour, while for the alternatives the opposite holds true. In transport, for example, the costs, travelling time and environmental pollution linked with car use are often trivialized, whereas long travelling times and lack of comfort in public transport and bicycles are emphasized (Rooijers 1992). Only when the dissonance on the current situation and behaviour becomes so large that it is impossible to deny it, may the behaviour of individuals change.

Individual preferences and attitudes. Besides these cognitive factors, motivational and even emotional factors also play a significant role. Such factors include such aspects as pleasure, privacy, status, personal control and representativeness. In transport, these factors may largely contribute to the preference for individual travel modes.

Social and personal values. People do not live in isolation, but act in a social context, which may have a substantial influence on the decision-making process and the behaviour of a person. In fact, people are strongly motivated to compare themselves with other relevant groups and to conform to general trends in a social context (Festinger 1957). Here, too, it should be observed that especially individually perceived social norms are important, which differ from objective standards. So the result is that the perceived social norms are distorted or influenced by individual preferences and attitudes.

Not only social but also personal norms may be relevant as well. These are the individual's behavioural motives, which are of importance for himself, but which are more or less isolated from individual preferences and perceived social norms. For instance, an individual may have a strong preference for car use, but nevertheless he chooses to use his bicycle for health reasons.

So the existence of habits hampers the possibility of changing behaviour. To break out of undesirable habitual behaviour, the individual will first need to have knowledge about all relevant alternatives (Ronis et al. 1989). Subsequently, the decision-making process itself can be influenced by control measures that affect the factors mentioned previously.

Consequences for transport. It is thus clear that there is a bias in the behaviour of individuals, which is rather conservatively oriented. This is a major obstacle to the achievement of a sustainable transport system (relationship 1 in Figure 12.1). Both the travel need and the travel mode choice are important in this respect; the first one is not discussed in this paper, although especially land use and residential policies may structurally reduce travel needs. It may also be concluded that the freedom in the transport decision-making process is mainly limited to the selection of travel mode in respect of new and incidental trips. Were this process also to take into account repeated similar trips, positive experiences (or absence of negative experiences) with a certain travel mode, this acts as a positive reinforcement of the choice made before, which increases the chance that the same mode will be chosen on the next occasion (Rooijers and Steg 1991). This may have at least two important consequences:

- there is hardly an explicit choice; the travel mode is used because it was used before, and therefore other alternatives are not even considered. In this respect the cost structure of car use vs. alternatives plays a large role, too: the marginal costs of car use – once a car is bought – are low; the marginal costs of public transport are much larger; alternatives become very expensive;

- persistent prejudices and misperceptions may develop which can hardly be remedied.

It should be acknowledged that the chance of habit formation is probably the highest for car use, because the car is more multi-functional than other travel modes; therefore, it has a highly attractive value; as a result, car use will not quickly lead to negative experiences.

In conclusion, because of the mechanisms described above individuals are biased towards their previous behaviour; few incentives to change behaviour, and it will be very hard for public transport and/or new technologies to gain a large(r) market share in the transport sector.

As will be discussed in Subsection 12.2.4, the influence of individuals on several pressure groups – especially the car industry and environmental groups – may be rather high (relationship 2 in Figure 12.1). The first is dependent on the demand in the market, while the second may reduce dissonance of individuals. However, in practice this potential still has to be mobilized. The above observed psychological bias may hamper this mobilization. The same holds for the influence of individuals on politicians (relationship 3 in Figure 12.1), which will be discussed now.

12.2.2 The government and sustainable transport

The next important stakeholder is the government (relationship 4 in Figure 12.1). Governments may influence the achievement of sustainable transport largely by introducing regulations, imposing taxes, providing subsidies, etc. In theory, the government should aim in this way at a socially optimal result in which the environmental impacts have to be weighed against economic (monetary) gains.

At present, devolution trends seem to dominate governmental policies. There are clear trends towards privatization, deregulation and governments 'stepping back', e.g., by reducing social benefits, subsidies, etc. Market forces seem to become more dominant in all regulated sectors. One of the well-known examples in transport are the policies to make railway companies more independent of governments; in some countries railways may even be privatized (e.g. in the United Kingdom). It may be concluded that also the government is increasingly subject to market forces and is increasingly rationalized on the basis of efficiency principles (Self 1993).

In public choice theories it is taken for granted that governments are composed of people, who also have their own indigenous objectives and utility functions. Important in this respect is that mostly a distinction is made between politicians (and political parties) on the one hand, and civil servants on the other. This distinction may be necessary because of the different utility functions both groups may have. Both groups will be discussed below.

Politicians and sustainable transport. In public choice theory it is assumed that politicians are in the first place focused on re-election (Frey 1983). Only when this is more or less secured, may ideologies and altruism come into play. Certainly in the present time, when voters seem to shift very easily from one party to another, re-election receives more emphasis than in the past; this reduces the possibilities for politicians to act in an altruistic way, e.g. in order to stimulate a sustainable transport policy stronger than the voters desire (relationship 3 in Figure 12.1). As a consequence, it may be expected that politicians will carefully monitor and follow the opinions of their voters. For sustainable transport again, the inherent conservatism of citizens is relevant for the behaviour and span of control of politicians.

At the same time, one should be aware of the large shifts in voter opinions. As observed by Rietveld (1997), the priorities of voters may drastically change, depending on such things as general economic conditions, unexpected disasters or threats, etc. For example, in the Dutch election in the year 1989 environmental issues had a high priority in the view of voters; in the year 1986 these issues had a much lower priority. Also in recent years, environmental issues seem to receive less attention.

Particularly important in this respect is thus the voting behaviour of individuals. During election periods people may vote for parties which promise to care for social and environmental issues. However, when measures are introduced which influence the life of the voter, these measures are often strongly opposed. As a result, politicians may act in a different way during election time than after the elections. One of the basic assumptions may therefore be, that 'political parties formulate policies in order to win the elections, rather than winning elections in order to formulate policies' (Downs 1957). A result of this observation is that politicians often remain vague during election times, promise too much (only to some extent: credibility is also important), and try to bind the median voters (see e.g., Mueller 1989).

Another important observation is that voters may choose a party because of many relevant subjects involved. It is likely that one party will have a preference for a given subject (e.g., financial/budgetary policy), while another party is preferred for other reasons (e.g., environment). A voter has to make a trade-off, and will vote for the party with the optimal mixture. A result, however, may be that the preferences of all voters are not consistent, as is witnessed by the Arrow paradox (A > B, B > C, A > C; see also Downs 1957). This provides a politician some degrees of freedom, which however may be used for personal gains. Additional objectives of politicians which are often mentioned in the literature are: getting monetary benefits, increasing personal power, receiving a place in history (books), altruism, ideology, etc.

An important feature of many environmental issues (especially climate change problems) are the long term impacts and the wide range of solutions that are necessary. Because of the short term in which elections take place, such long term problems tend to be more or less neglected by politicians. The main reason is that it is not possible for politicians to present results of their policies before the next elections, while at the same time there are many negative impacts for their voters (e.g., restrictions in car use). This trend may be reinforced by the increasing impact of the media in past decades. Because individuals can more easily be reached and informed, the degrees of freedom for politicians are decreasing even more. In this respect also the policy cycle as presented by Van Dijk (1991) may be illustrative. This cycle is as follows:

- the environment becomes a political issue (e.g., because of a disaster, new information);
- the issue is taken over by politicians and it becomes an electoral issue, therefore many policy measures are promised;
- after the elections concrete targets and instruments are to be introduced: an 'information battle' starts, in which lobby groups (e.g., environmental groups, industries, etc.) inform the policy maker. As will be noted in Subsection 12.2.4 environmental groups may be weaker in this respect than their competitors; and
- new decisions are taken; however, in the previous stage many possibilities are not regarded as acceptable, while later on the issue may not be that relevant anymore in politics; therefore, strong measures are (often) not introduced.

Civil servants and sustainable transport. Civil servants will normally have a utility function, which differs from the societal or the political one; this may sometimes lead to a suboptimal allocation of funds because of reasons of bureaucratic power or self-esteem (Nijkamp and Rienstra 1997). An example of such a utility function is discussed in the budget maximization theory (see e.g., Dunleavy 1991), which takes for granted that the utility function of civil servants correlates positively with the public budget he has at his disposal. Since the civil servant has a monopoly position in the provision of information to the parliament, he will supply information with the intention that the intervention level is higher than in the societal optimal situation. This is illustrated by a graphical presentation in Figure 12.2, in which a simple situation is presented with linear cost and revenue curves and without fixed costs. In this figure the level of intervention is supposed to be related to the budget of the civil servant: a budget of a civil servant increases when the intervention level increases.

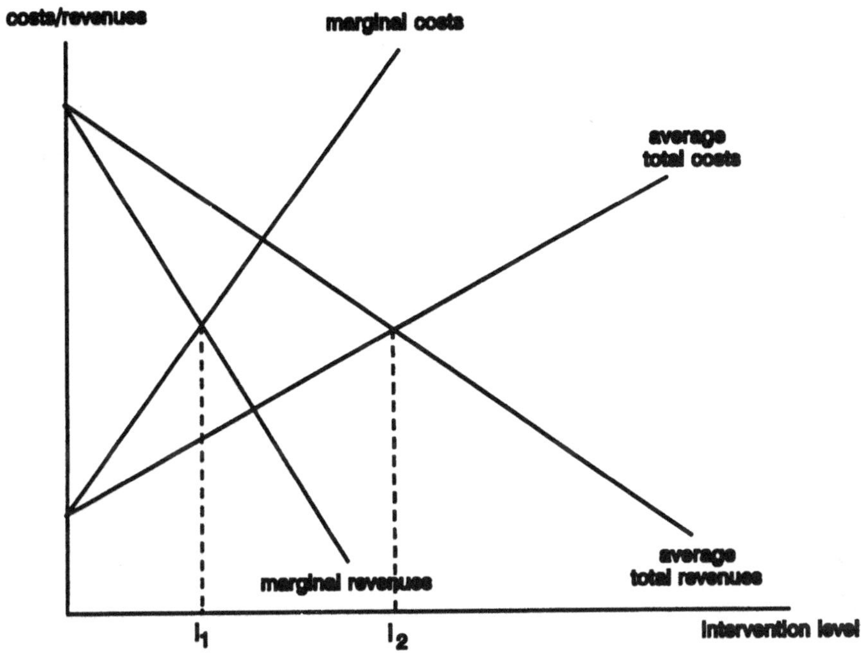

Figure 12.2 The intervention level under a budget maximizing civil servant.

Were the parliament to have full information, it would choose the intervention level at which the marginal costs would equal the marginal revenues, which corresponds to an intervention level of i_1, which is optimal. A civil servant, however, may provide only information about the total costs, and as a result the parliament may decide to intervene at (somewhat to the left of) the point where the average total costs equal the average total revenues, which corresponds to intervention level i_2. This is optimal for the civil servant since it maximises his budget, but it is suboptimal from a societal point of view. In this respect it is also important to notice that the costs of receiving information from third (neutral) parties is high in most cases. There are also other theories, like the bureau shaping theory (Dunleavy 1991). In this theory higher ranked civil servants are not focused on budget maximising, but on status, prestige, patronage and influence of their work tasks. This objective may not correspond with maximizing budgets, but also not with the achievement of a socially optimal result. In general, it may be concluded, however, that the objectives of civil servants do not necessarily have to fit with attempts to achieve a social optimum. In theory, there are also other circumstances which increase the influence of civil servants, but which may not result in a social optimum (Frey 1983):

- in the public sector there is little incentive to act in a socially optimal way; a civil servant is seldom confronted with the consequences of decisions, especially when he acted in the formally right way;
- it is difficult to measure and evaluate the output;
- there is no clear market demand for their publicly offered services, because mostly public goods are provided;
- citizens avoid problems with civil servants, because they need them to grant permits, etc.;
- in the public sector there are often limited possibilities for extra wage increases, promotions, etc., so that there may be few incentives for a good performance.

This attitude may result in suboptimal solutions, in which the governmental influence becomes too great, while changes in strategies and policies are not encouraged. In transport, this may result in too large an intervention in the transport system. Some consequences may be:

- the price asked for using the infrastructure may be too low, e.g., to satisfy car users (who form a powerful pressure group) and to maximize subsidies (budgets);
- public transport companies are protected too much, because then the influence of civil servants is the highest;
- new technologies and ideas are not introduced, in order to avoid failures.

In practice the notion of one group of civil servants may be too simple. Several departments usually have their own preferences and utility functions, e.g. a department of Transport or Economic Affairs is usually more in favour of constructing new (road) infrastructure than a department of Environmental Affairs. As a result, long negotiations and 'information battles' take place between various departments and interest groups, resulting in even less efficient decision making.

One of the conclusions may anyway be that the influence of individuals on the behaviour of civil servants may be expected to be small, as is also demonstrated in Figure 12.1. The utility function of civil servants may result in suboptimal solutions and hamper the creation of new policy strategies and solutions.

12.2.3 International organizations and sustainable transport

From a public choice point of view, countries tend to join an international organization (as a club) when (Vaubel 1991):

- international externalities occur, which result in an underproduction of international public goods (e.g., peace keeping) and/or an overexploitation of common resources (e.g., emissions of greenhouse gases; water supply in certain areas);
- international economies of scale occur in the production of public goods (e.g., international standards for certain goods; encouragement of R & D of certain industries);
- cooperation is beneficial for all parties; non-cooperative behaviour produces suboptimal outcomes ('prisoners dilemma') and cooperative behaviour improves the outcome (e.g., international trade agreements, environmental measures); free riding is a main problem in this case.

International organizations become increasingly important for all kinds of policies. This is the result of the strong globalization of the world economy; markets are opened and liberalized, requirements for products are standardized, etc. In transport this results in an increasing cooperation of countries, e.g., in the construction of an HST network and other Trans European Networks. This increasing importance is especially the case in Europe, where the European Union gains increasing influence in policies, while national governments lose autonomy. As a result, environmental measures are also taken by international organizations, in order to prevent 'free rider' behaviour by certain countries. This trend is also observed in the transport market (relationship 7 in Figure 12.1). Several examples can be given of the decreased freedom of national governments in the transport market:

- fuel prices cannot be determined in an independent way, because people may buy fuel 'on the other side of the border';
- protection of the railways is abolished, because of new European regulations;
- companies are becoming increasingly 'footloose': when environmental measures are taken in one country, companies invest in other countries where the resulting costs are lower; a well-known example are the discussions in several countries about the introduction of a CO_2 tax;
- in an integrating economy like the EU, certain measures are often forbidden, since they may hamper free trade principles.

It may be concluded that many measures have to be taken within international organizations. This holds especially for externalities which an impact internationally (e.g., the greenhouse effect) and for international transport modes (e.g., aviation). At the same time, one country can often hamper the introduction of measures because of voting procedures within these organizations. As a consequence, many independent issues are also politically linked to each other, which hampers the effective introduction of new policies. This inertia is reinforced by the often inadequate democratic control: media are in general nationally-oriented, while the democratic mechanisms do not work within international organizations, and especially individuals seem to have little interest and influence in these organizations. Therefore, policies of international organizations are prepared in a 'black box', in which lobbies of individual countries and pressure groups operate fruitfully.

12.2.4 Pressure groups and sustainable transport

The final group of stakeholders identified in the conceptual model is made up of pressure groups, which may have a far-reaching influence on policies of both governments and international organizations (relationships 6 and 8 in Figure 12.1). At the same time individuals influence the existence, behaviour and impact of pressure groups (relationship 2).

Pressure groups tend only to come into existence when collective efforts provide certain advantages to the members of a group. Their activity can be considered as a public good; therefore, there are little incentives to become a member. There are three conditions however, which may make a group stable and effective (Frey 1983):

- in small groups free riding becomes impossible because of social control; this makes small pressure groups often more effective than large ones;
- a group may provide a private good to the members; therefore, it becomes attractive to join the club. Well-known examples are mobile repair services for car drivers, an interesting journal, access to information, etc. However, also organizing social activities may be a reason to join, or the feeling to belong to an exclusive club;
- governments may force a group focusing on a specific issue into existence, because they prefer to deal with formal groups.

Individuals with common interests, but without sharing explicitly one of the above reasons, constitute 'latent groups'. These groups may only come into existence when policies interfere too much in a negative way. As a result, groups in society differ in power and strength: labour unions and employer organizations, for example, are well organized, while consumers and tax payers do not have powerful pressure groups. Also environmental lobbies may be relatively weak because of their large size and diffuse objectives. By negotiating, the most powerful groups may gain advantages at the expense of the less powerful ones. For each intervention measure, the costs per individual (consumer, tax payer) may be so low and unclear, that it is not rational to resist the measure (like minimum prices, protection measures, etc.).

Strong groups may have great power in a decision-making process (Mueller 1989). Governments often depend on information provided by the groups, while in exchange they may receive services from the public sector. They also may influence voter groups or mobilize people.

As observed above, the influence of pressure groups in international organizations may be large, in particular because of the lower democratic control and the low participation of individuals in the process (relationship 8 in Figure 12.1). Also the efficiency of lobbies of national governments seems to be very important in the decision-making process of international organizations (relationship 5). Several groups may be identified which influence the transport sector to a large extent (relationship 9): the car (and aviation) industry, railway companies,associations of car drivers, the oil industry, environmental groups; these will now be discussed briefly.

The car (and aviation) industry. The car industry is first focused on maximizing profits and on the protection of its market share. In the end, this industry often depends on the demand in the market, i.e. on individuals. Individuals are clients of either the car industry, the railways or other transport suppliers in order to satisfy their travel demand. In a free market, one may assume that supply is determined by demand of individuals, so that the car industry would follow the demand of citizens. For the reasons mentioned in previous sections, however, citizens seem to be very conservative in their attitude towards changes, and therefore the car industry seems to have little incentive for developing more sustainable cars and transport technologies.

In the past decades, a strong concentration has taken place in the car industry. As a result, very large companies are now present in many countries (e.g., Germany, France, Japan, etc.), which provide much employment and form an important economic sector. As a result, these companies have a strong influence on government policies at both the national and the international level. For example, the European car industry which is facing an increasing competition from Japanese and American car industries has – as a result of their large economic importance for the European market – been heavily protected. Another example of the influence of this industry can be found in the tax credits given in France and Spain when people buy a new car; this measure was taken in order to stimulate employment in the car industry. A main aim of these industries may be to protect their investments once made, and to reduce costly investments in new technologies which are not directly demanded for in the market. As a result, investments in environmentally friendly technologies may be low.

Because our scenarios in Section 12.3 do not deal explicitly with aviation, we will not pay much attention to these pressure groups. Similar behaviour as in the car industry is likely to occur, however; both airports and airline companies will try to maximize growth and have little incentive to reduce external impacts. Since this transport mode is international in nature, it is very hard to impose policy measures because of free rider behaviour of countries (also because this rapidly growing industry is regarded as very important for economic and employment growth). This may be a reason that no taxes have yet been levied on kerosine and tickets.

Railway companies. A second important player at the interface of transport and environment is formed by the railway (and other public transport) companies. Governments have invested largely in rail infrastructure and have usually covered the operating losses of these companies, while – as a result of ownership conditions – governments were also otherwise highly involved in these companies. Rail companies seem to have a strong lobby at the national level, because stimulating public transport is a main objective of most governments in order to reduce congestion and environmental impacts. Also the social task of public transport – providing mobility facilities to people who cannot afford a car, elderly, remote regions, etc. – makes this sector important for government policies. Finally, the employment in this sector is also considerable, while the employees in this sector are often highly organized in labour unions and are also very willing to strike, which may have strong economic and social impacts.

Associations of car drivers. A third influential group is formed by associations of car drivers. Because of the strong psychological impact of car driving, strong anti-car measures are very unpopular and opposed by strong lobby groups in all countries. These pressure groups have succeeded in recruiting a large membership, e.g. by providing mobile repair services and popular magazines. As a consequence, the influence of these groups may be extremely large, also because they are able to mobilize many people. Restrictions on car use are therefore unpopular in every western country.

The oil industry. Another important pressure group may be the oil industry. In the first place, this sector employs many people, but also the financial power of this industry is extremely great. It is obvious that this industry favours the use of fossil fuels to the maximum extent, so that reductions in oil use may be opposed by these industries. Because of the international character of the industry, the influence of this sector on international organizations may also be very large.

Environmental pressure groups. A final important group are environmental pressure groups. As discussed before, these may be relatively weak, because they have to combine opinions of a lot of people, which may result in free rider behaviour and in a 'latent group'. An exception are local and regional groups, which may play an important role when a group of individuals is facing local externalities, like noise annoyance and visual intrusion. These groups are often relatively small, which makes free rider behaviour more difficult. At the same time the objectives of such a group are usually pronounced and clear, which makes the group stronger. Well-known examples are the so-called NIMBY (Not In My Backyard) pressure groups, which oppose large infrastructure projects. These groups operate often in a very effective way.

Environmental groups may also be important at the national level, however. Their influence may increase because of alarming information provided by scientists (e.g., the Club of Rome, the recent IPCC-report) or by disasters (depletion of the ozone layer, floods), but also because of their own field work (e.g., noise and stench nuisance, visual intrusion). As argued above, individuals often support such groups in order to reduce their dissonance. These groups also may gain support by providing information to the public. The 'marketing' of such a group therefore becomes very important (e.g., Greenpeace). However, the environmental groups are mostly considered to be weaker than their opponents, not only because of differences in size, but also because they try to achieve improvements in the long run, while politicians are often focused on the short term (e.g., employment in the car industry).

From our conceptual model it can be concluded that most forces have a negative impact on the attempts to achieve more sustainability. Most stakeholders in the decision-making process seem actually to be in favour of the current situation, while changes in the system and society are not popular and are not shared by large majorities.

Next, we will test our conceptual model in an empirical way, by presenting various results on expected and desired developments in many fields that influence transport and the transport system itself.

12.3 An Empirical Test: From the Expected to the Desired Future of Transport

For the empirical investigation of the future of transport reported in this paper, a postal survey questionnaire was sent to Dutch transportation experts. In a second round, an earlier version of both scenarios was validated and commented on by the respondents from the first round as well as by a sample of European transport experts.

The reference year of this experiment is the year 2030; the questions concern the Western European passenger transport system. For a detailed description and accounting of the questionnaire used we refer to Nijkamp et al. (1997); here we only note that the response rate was 36% (n = 271), and that the response gave a representative picture of the sample.

In the questionnaire respondents were first asked to indicate both expected and desired future developments of background factors influencing transport, which can be found in the spatial, institutional, economic and socio-psychological field. Next, the expectations and desires on the use of several conventional and newly developed transport modes were investigated. In this way it is possible to construct an expected and a desired scenario, while the discrepancy between reality and wish can also be investigated. Here we will first briefly describe the results of this experiment. Next, we will analyse the role of the stakeholders in more detail.

12.3.1 Outline of the expected and desired scenario

The main developments and characteristics of the expected and desired scenario are presented in Table 12.1; for a more extensive elaboration of the scenarios we refer to Nijkamp et al. (1997). In the experiment it is taken for granted that transport demand is mainly a derived demand, influenced by the above mentioned factors. The resulting transport system is presented in Table 12.2.

	Expected Scenario	**Desired Scenario**
Spatial	* Concentration of population and activities in European economic core zone * Moderately compact city introduced	* Peripheral regions develop positively * Compact city introduced
Institutional	* No clear policy-orientation toward sustainable transport * No clear separation of public and private responsibilities (railways, infrastructure)	* Regulatory policies are widely introduced * Infrastructure, railways etc. are publicly owned
Economic	* Transport system not very profitable * No clear policy level which is most important	* Transport system is rather profitable * Much centralization in economic policies (EU)
Socio-psychological	* Individualization trend continues * Little attention for equity, both in society and in	* Individualization trend reverses * Much attention for equity, also in transport

	transport	

Table 12.1 Main characteristics of the expected and desired scenario

Source:Nijkamp et al. (1997)

From these tables it can be concluded that in the expected scenario various trends and underlying factors will not change to any great extent. For example, no large scale changes are expected for individual citizens, who are not expected to modify their behaviour drastically in order to reduce the externalities caused by transport. The same holds for the government, whose policy is not expected to be sufficiently and clearly focused in one direction. Especially future spatial developments appear to be a focal point of attention of the various transportation experts in the panel.

In the desired scenario a clear choice is made for a collective transport system. Therefore, large scale changes are necessary in the behaviour of individuals, as well as in the institutional and economic environment. At the same time the government should invest largely in infrastructure of collective modes and take measures to reduce car use to a large extent. In this way the policy is clearly focused on regulation of the transport sector and on the stimulation of a modal shift towards collective modes.

As demonstrated by Rienstra et al. (1995), the expected scenario will not result in large scale reductions of the use of conventional fuels and CO_2 emissions; the desired scenario however, might result in reductions – depending on several assumptions – of about 80%, which may be sufficient to achieve a 'sustainable' emission level.

	Expected scenario	**Desired scenario**
Long distance	* conventional private cars dominate modal split * large scale introduction HST * collective modes offer supplementary service	* Collective modes dominate (HST, trains) * Conventional fuels largely replaced by new fuels (electricity, liquid hydrogen)
Urban	* More introduction of metro and light rail * Introduction electric cars * Conventional cars most important	* Collective modes dominate * Car use heavily restricted * Walking and cycling increases

Table 12.2 The expected and desired long-distance and urban transport system

Source:Nijkamp et al. (1997)

12.3.2 The role of the stakeholders in the expected scenario

It is clear that there are large discrepancies between the expected and desired future of passenger transport. It is now interesting to analyse whether the stakeholders identified in the conceptual model of Section 12.2 behave in the way assumed, resulting in a system which is not expected to move to sustainability (as is indicated by about 80% of the respondents).

Individuals. One of the striking features of the expected scenario is that individuals are not prepared to make sacrifices for a sustainable transport system. In general, they are not expected to change their behaviour, e.g. by travelling by public instead of private modes, by living in compact cities and by a reversal of the individualization trend. These trends may be regarded as an extrapolation of past trends, and may be explained by cognitive dissonance theories (see Subsection 12.2.1).

The government. The government in the expected scenario is rather ambiguous. All kinds of measures are taken, but there seems to be no clear policy focus. Car use is to some extent restricted by road pricing, parking measures, etc. These measures are not introduced in an adequate way, however. The policy also fails in providing sufficient alternatives by large scale investments in collective modes and their infrastructure, while the responsibility (public, private) for the provision of these services is not well organized, either.

The only policy which is expected to be introduced is a rather strict spatial policy at both the urban and European level, aiming at geographical concentration and a compact urban spatial structure. Such a policy reduces travel distances and creates voluminous travel demand per trip, and as a result collective modes become more competitive. Apparently, there will be a consensus in society on these issues, which is rather striking since the living conditions in a compact city may not always be very preferable. It may be the case that nature parks etc. become increasingly important in the opinion of individuals, while civil servants will tend to like the regulation of the spatial structure.

International organizations. International organizations are not expected to dominate the political scene. In the area of economic and transport policy, for example, the European Union is not expected to become the most important authority. This makes it difficult to coordinate policies and to reduce international externalities. In this way free rider behaviour of individual countries becomes an attractive option. This may explain why several measures – like road pricing, increasing fuel prices, restricting car use – may not be introduced to a sufficient extent, because it affects the competitive position of individual countries. At the same time, this hampers the coordination of the construction of a network of collective modes, like the HST network. An advantage may be that the decision-making process takes less time and that fewer compromises are necessary, but this is not expected to compensate for the disadvantages.

Pressure groups. Pressure groups, finally, seem to behave as is assumed in the conceptual model. Therefore, individual car use is still dominant, while few changes are introduced in the transport system. The car and oil industry succeed in reducing the dissonance of individuals, and influence the trade off between short-term (economic) and long-term (environmental) issues within national and international authorities. Environmental groups will apparently not be successful in changing these opinions, as is explained by the conceptual model. The same holds for railway companies, who are not able to gain a larger market share.

In conclusion, the conceptual model presented in Section 12.2 appears to offer a reasonable mapping of expected real-world developments. The main stakeholders in the process seem to behave in the way predicted in this model. As a result, the system is not expected to change in the direction of more sustainability. Next, it is interesting to investigate how the stakeholders should act in order to achieve a more sustainable transport system, according to the Dutch transportation experts in our sample.

12.3.3 The role of the stakeholders in the desired scenario

In the desired scenario a clear choice is made for regulation and the encouragement of collective transport modes. This requires an entirely different behaviour of distinct stakeholders.

Individuals. Individuals may be the starting point for the necessary changes, because in a democratic society with an economy based on free market principles, both the government and industry will finally follow these opinions. In the desired scenario the individuals are more or less forced – but also prepared – to reduce car use and to use collective modes or to shift to cycling and walking. Such a behaviour can only be achieved when the dissonance of individuals becomes so large that they change their behaviour. This may be achieved by the provision of information in a way which cannot be denied; another possibility is that the negative consequences of the current mobility behaviour become so clear (e.g., by a disaster) that individuals will change their behaviour. An important remark in this respect is that there should be alternatives available for current car use; large scale investments in collective modes are therefore necessary.

The government. The national government should – in the desired scenario – shift responsibilities to regions and international organizations (regionalization and centralization); therefore the influence of this government level will decrease. International externalities should be coped with by international organizations, while local and regional externalities might be coped with by regional and local authorities. In this way the subsidiarity principle may also hold for the transport policy. When this is achieved, free rider behaviour of national governments can be avoided, because relevant measures are compulsory. The task of governments is to focus their policies clearly on the provision of collective modes and the reduction of car use. Large scale investments are necessary in the infrastructure of these modes. In the urban areas also a policy focused on the compact city may be a key success factor for the achievement of the desired direction.

It is clear that in this situation both national politicians and civil servants should be prepared to give up part of their authority, while at present this level is very influential. It is obvious that this will cause many problems in practice. These may first be found in the utility functions of politicians, which have to support this shift. They also should be more focused on long term sustainability issues instead of short term electoral success.

In the government, too, incentives should be introduced for achieving clear targets for civil servants, e.g. by setting a clear time path which can easily be checked by the media and individuals. Also incentives (e.g., monetary, promotion, etc.) can be given to civil servants, in order to change their utility functions. In this way regulation should not be reduced, but made much more effective in order to serve the achievement of general sustainability objectives and a societal optimal result.

International organizations. As mentioned above, international organizations should become more important, while also some competence should shift to regional authorities. In this way externalities may be coped with at the level where they occur. It is clear that several measures can be introduced much more easily than at the national level (road pricing, increase in fuel prices, cooperation of railway companies). However, it is necessary that the decision-making process within these organizations will change. National interest should not be a main issue in this process, therefore these organizations should become more independent of national governments. Therefore, an authority like the EU should gain more independence; the decision-making process should become faster and more efficient. At the same time, however, there should be more democratic control, since otherwise the decision-making process will be steered by pressure groups and the civil servants of these organizations. Instead, individuals – with opinions as discussed before – should steer these organizations. When this is not achieved the legitimacy of these organizations is at stake and policy measures will not be accepted in society.

Pressure groups. Pressure groups, finally, should largely change their attitude. The car industry mainly follows the demand of individuals. As a result of this scenario, the car industry will largely diminish, while new technology will be based on cars powered by electricity or other alternative fuels. Therefore, this industry should not protect its current position, but develop new markets and strategies; for example, by shifting funds to the development and promotion of 'zero-emission' cars. An entirely new market may be found in the development of new collective modes and the production of these modes. If the desired scenario is to become reality, a lot of production capacity and funds are needed and a large market may be found in this field. This capacity may largely be found in the car industry.

The oil industry may lose its position to an even larger extent, since the use of fossil fuels may largely diminish. However, also in the future there will continue to be a high demand for energy in order to power the transport system. This industry should therefore focus on the development and production of new sustainable energy forms.

Railway companies and urban transit authorities should change their attitude from defending their protected position and subsidies towards a market-oriented organization. In the desired scenario their market potential is extremely high; the railway companies should therefore be much more eager to meet this challenge. As a result, more flexibility, efficiency and market orientation is a *sine qua non* for the success of such a policy. In this way also the profitability of this system should largely increase.

Environmental pressure groups, finally, may play a vital role in making society more conscious about the environmental consequences of mobility and in creating dissonance in society. These groups should therefore carefully present their information to individuals, the government and international organizations. Of course, the credibility and public acceptance of these organizations is important.

12.4 Conclusions

From the conceptual model developed in Section 12.2 it may be concluded that most forces in the decision-making process hamper the development of policies clearly focused on sustainability in the transport sector. This observation also holds when the expectations of transportation experts are analysed: most trends are expected to continue and will not change to any great extent. In the desired future, however, the role of the stakeholders differs largely from the expected future. If this scenario were to become reality, a more sustainable future may be achieved.

Clearly, the current decision-making process will not stimulate the transformation towards a sustainable transport system, although this need and the related objective is acknowledged by a large majority in society. Individuals may – and in a democratic society with an economy based on free market principles should – be the starting point of possible changes. At least three key success factors may be identified.

First, it is necessary that the information about the external effects of the current mobility behaviour becomes so appealing that it creates a dissonance that is so large that individuals want to change their behaviour. Environmental groups, scientists, and national and international authorities may play an important role in the provision of this information. When this occurs, governments cannot deny this trend, and measures may have strong support in society. New ways of information provision may have to be found, therefore. In the end the (car and oil) industry will have to follow the demand of their market and offer such alternatives in order to survive.

A second key success factor is the presence of appealing alternatives for the current mobility behaviour (new technologies, collective transport modes) (see also Rienstra et al. 1996). If these are not available, individuals are not able to change their behaviour and attitude, even if they want to do so. The government should invest heavily in the infrastructure of collective modes and bicycles, so that alternatives become available; a clear policy for stimulating such a shift is a necessary condition too. Railways should be prepared to change their organization and provide an alternative to the current car system, in cooperation with other providers of collective transport (urban public transport, bus companies, taxi companies). Finally, the car and oil industry should invest in alternatives and have to find new markets and strategies.

The third success factor however, may be most important. All stakeholders need to change their attitude and cooperate in order to cope with the challenge of making the radical shift towards a sustainable society. This may be achieved by changing incentives in the conceptual model pointed out; there should however become a willingness to change these – e.g. by acknowledging the failure of the current way of decision-making – otherwise such a change is not feasible. New policy options and strategies may then be identified and introduced, and a sustainable transport system may become a realistic and deliberate option.

References

Brady GC, Clark JR, Davis WC. The political economy of dissonance. Public Choice 1995; 82:37-51.

Coase RH. *The Firm, the Market and the Law*. Chicago: The University of Chicago Press, 1988.

Dijk, F van. 'Decision making about the environment: the role of information'. In *Environmental Protection: Public or Private Choice*, Kraan DJ, In 't Veld RJ. Dordrecht: Kluwer, 1991.

Downs A. *An Economic Theory of Democracy*. New York: Harper and Row, 1957.

Dunleavy P. *Democracy, Bureaucracy & Public Choice*. New Jersey: Prentice Hall.

Festinger L. A theory of social comparison processes. Human Relations 1954; 7:117-140.

Festinger L. *A Theory of Cognitive Dissonance*. Evanston: Peterson, 1957.

Fishbein M, Ajzen I. *Belief, Attitude, Intention, and Behavior: An Introduction to Theory and Research*, Reading: Addison-Wesley, 1975.

Frey BS. *Democratic Economic Policy: A Theoretical Introduction*. Oxford, Basil Blackwell, 1983.

Mueller DC. *Public Choice II*. Cambridge: Cambridge University Press, 1989.

Nijkamp P, Rienstra SA. 'The public private nexus in financing infrastructure investment'. In *European Transport Networks*, Button K, Nijkamp P, Priemus H., eds. Aldershot: Edward Elgar, 1997 (forthcoming).

Nijkamp P, Rienstra SA, Vleugel JM. *Transportation Planning and the Future*. Chichester: John Wiley, 1997 (forthcoming).

Rienstra SA, Smokers RTM, Nijkamp P. Energy scenarios for european passenger transport in the year 2030. TI-discussion paper 5-95-244. Amsterdam: Tinbergen Institute, 1995.

Rienstra SA, Vleugel JM, Nijkamp P. Options for sustainable passenger transport; an assessment of policy choices. Transportation Planning and Technology 1996; 19:221-233.

Rietveld P. Political economy issues of environmentally friendly transport policies. International Journal of Environment and Technology 1997; 7:398-416.

Ronis DL, Yates JF, Kirscht JP. 'Attitudes, decisions, and habits as determinants of repeated behavior'. In *Attitude Structure and Function*, Pratkanis AR, Breckler SJ, Greenwald AG, eds. Hillsdale: Lawrence Erlbaum Associates, 1989.

Rooijers AJ. *Meningen en Motieven ten Aanzien van Autogebruik voor Woon/Werkverkeer*. VK 91-14. Haren: Verkeerskundig Studiecentrum (VSC), 1992.

Rooijers AJ, Steg EM. *De Rol van Gewoontegedrag bij Vervoermiddelkeuze*. Haren: Verkeerskundig Studiecentrum (VSC), 1991.

Self P. *Government by the Market; The Politics of Public Choice*. Basingstoke: MacMillan, 1993.

Vaubel R. 'A public choice view of international organization'. In *The Political Economy of International Organizations; A Public Choice Approach*, Vaubel R, Wilett, TD. Boulder: Westview Press, 1991.

Verhoef ET. *The Economics of Regulating Transport*. Cheltenham: Edgar Elgar, 1996.

13 SUSTAINABLE DEVELOPMENT AND SOCIAL JUSTICE: THE TOOL OF THE REDUCTION IN INCOME INEQUALITIES

Jean-Marie HARRIBEY

Introduction

Based on Georgescu-Roegen's works on entropy and Rawls's works on social justice, a double theoretical reflection is essential, showing that a long-term development strategy that is both environmentally and socially sustainable, is only meaningful as long as productivity gains are primarily used to generate an everlasting production and consumption growth but in order to reduce all individals' working time. We propose a link between the concepts of equity and solidarity both between and within generations.

In order to give our approach a theoretical basis, we propose a simple mathematical model for the reduction in the income scale, making an unemployment decrease possible while protecting ecological balances from the threats of excessive economic growth. Then the model can function in two ways. First, for a given unemployment rate, it allows either the determination of the scale of reductions in inequalities required to cut down unemployment within a given period. Second, the time necessary to reduce unemployment can be calculated according to a decision about the socially desirable and/or sustainable income hierarchy.

We test our model on the French economy by following three stages. First we apply it to the current distribution of household incomes. Then we carry out the same type of work on wage distribution alone. In both cases we show that unemployment elimination is a distributional problem and not one due to wealth level. Finally we perform a third simulation in order to reduce the consequences in terms of consumption growth of a too high rise of low and average incomes, whose marginal propensity to consume is higher than that of people whose incomes are penalized by the reduction of inequalities. We reduce the period of time necessary to cut down unemployment, and the improvement of the standard of living of the lower classes depends to a small extent on raising their purchasing power and to a large extent on shortening working hours.

In conclusion, we indicate that instead of suffering from the effects of a moderate growth, it is possible to take advantage of this situation to trigger off social and cultural changes in the ways we consider for the improvement of welfare.

At the end of the 20th century, development is encountering two limits: the first is environmental, because there is no endless economic growth that is materially possible; the second is social and cultural because, both on the global level and within industrialized countries, such phenomena as poverty, unemployment, inequalities and acculturation are growing in a worrying way. Economic growth should, it is said, reduce the scarcity which is hitting humanity like a plague. But it seems that, far from the reducing it, it is renewing it by forever extending the limits of need. Moreover, the logic of accumulation of capital encourages selective satisfaction of these needs by maintaining deep social injustices. In undeveloped countries, the majority of the population are excluded from the economic world, having seen the annihilation of their culture and traditions which gave a meaning to their way of life. At first, in developed countries, it was the least qualified wage-earners who suffered the consequences of the redeployment of world scale productive activities, but progressively all the remaining salaried groups are undergoing the same phenomenon. Thus, belief in an economic development, which will automatically create social consequences of a positive nature (trickle down effect), can be seen today as an ill-founded ideology, whose 'raison d'être' is to make us believe in an improvement in general material wellbeing while leaving the relative place of individuals and social groups unchanged. To expect the return of high economic growth to solve phenomena such as poverty and unemployment is to overlook the problem of distribution and to leave the reduction in inequalities solely to the distribution of the mythical fruits of growth, when in fact the structure of total product distribution would remain almost unchanged.

In the same way, the environmental crisis, laxity over major risks (nuclear energy, the greenhouse effect, climatic changes), the dwindling of vital non-contaminated resources (water, air), the programmed depletion of fossil fuel energy supplies, the extinction of certain animal and vegetable species (F.A.O., 1993) make the serious realization of sustainability an urgent matter. But what sort of sustainability? That of development? If it were a question of assuring the durability of development which has existed for two centuries, it would be contrary to the 'principle of responsibility' (Jonas 1979). The folly of a seemingly rational economic system has been responsible, in the name of profitability, for the plundering of nature and the exclusion from society of a growing mass of individuals who have become unemployable both for productive and consumer ends.

Can we content ourselves then with confining theoretical reflexion to the narrow framework of the *weak/strong sustainability* alternative? Weak sustainability wholly retains the Promethean dream according to which it will always be possible, thanks to technical progress, to substitute depleted natural resources by produced capital. Strong sustainability refuses this dangerous gamble and subordinates the growth of economic activities to the possibility of preserving the supply of natural resources over the years. However, one must go a step further, because even in the second case, economic growth continues to be considered as the principal aim, admittedly subject to additional constraints, but which is not in itself called into question.

The credibility of the concept of sustainability therefore depends on the ability to reopen discussions in various directions on economic growth.

- Economic theory cannot ignore the work of Nicholas Georgescu-Roegen which shows that economic activity lies within a physical universe subject to the law of entropy. According to him, economic development is founded on inconsiderate use of accumulated world energy supplies over the years. " That which enters into the economic process consists of *valuable natural resources* and (...) that which is discharged consists of *valueless waste*. (...) From the thermodynamic point of view, energy matter which is absorbed into the economic process is in a state of *low entropy* and comes out in a state of *high entropy*" (Georgescu-Roegen 1995, p. 55).

- The accumulation of capital which is at the origin of development has provoked the falling apart of global society (Polanyi 1944). This phenomenon must not be seen as a simple parallel evolution of the economy and other forms of life in society, but as an ever widening gap which is becoming less and less bearable. Social unsustainability built on increased inequalities, exclusion and acculturation can be considered as a form of *entropy of capital* (Latouche 1986) symmetrical to the entropy of matter. Therefore the economic machine and the 'megamachine' (Latouche 1995) transform '*usable* or *free* energy' into '*unusable* or *tied* energy' (Georgescu-Roegen 1995, p. 56), and, at the same time, enslave man by forcing him into a task whose end result is market exchange while depriving him of his powers of establishing the level of his own needs and the quantity of effort

> that he is prepared to supply. Economic rationality is always pushing the limits of necessity further and further into the distance. The result is an irresistible development explosion inherent in economic rationality which capitalism has been able to free from all hindrances.

Environmental and social crises are linked. One must therefore strive for sustainability along these two lines. Environmental and social sustainability implies the reinstatement of the concept of 'enough' (Gorz 1988) and the adoption of a 'renunciation code' (Jonas 1990); that is to say, a progressive renunciation of economic growth in industrialized countries. Hence, this reversal of perspective places the question of the distribution of natural resources between the generations, incomes drawn from economic activity and the means to produce goods and services (work) into the bosom of the present generation.

This pre-eminence of the question of distribution of wealth means that the theory of sustainability only exists within a theory of justice. John Rawls (Rawls 1971) has renewed the terms in which the problem of social justice in the utilitarian and Pareto approaches was traditionally raised (or rather suppressed). Unable to define collective wellbeing other than from a sacrificial point of view, the utilitarian conception of justice came up against a contradiction (Dupuy 1992). Therefore, by reasoning under a 'veil of ignorance', it is possible, according to Rawls, to define a social contract instituting operating rules for a just society which would respect two principles, one of non-negotiable freedom and the other of difference, which would itself be subject to two conditions: chances must be equal and inequalities should only be allowed if they obtain better efficiency which would improve the conditions of the most needy. Rawls departs from utilitarism when he underlines the importance of basic assets which he calls *primary social assets* of natural origin (strength, intelligence) and stemming from social organization (rights, liberties, power) to insure equal chances, as opposed to assets which are merely useful. He does not accept a conception of social well-being arising from the increase in per capita consumption and a concept of equity between generations which would amount to an intertemporal preference.[1] But Rawls stops at the point where it would be

1 "Finally, the last stage is not characterized by great abundance. (...) It is possible for greather wealth not to be useless as a means of reaching several objectives; in fact, average income is perhaps not very high in absolute terms. Justice does not require the present generation to economize simply to enable future generations to be richer. Economizing becomes a condition whereby fair institutions and equal liberties can be fully achieved for everyone. To add a supplementary build up of capital can only serve others aims. It is a mistake to believe that a just and honest society should go hand in hand with a high standard of living. Man is in need of work with a meaning, work in open association with the framework of fairly based institutions. An abundance of wealth is not necessary in order to reach this goal. In fact, beyond a certain level, this

necessary to move on from the *definition* of a just society to a *method* of justice in a yet unjust society, a method which would impose a *categoric imperative* of the Kantian type (Bidet 1995). For this reason, the 'abusive' use (Meral 1996, p. 306) of Rawlsian principles within the utilitarian paradigm has been made easier. By reducing justice to a mere summing up of individual interets, the sharing out of produced wealth and resources downplayed while the principle that one must have endless economic growth, even though it is known to be impossible, hinders the struggle for equity between the generations (Norton 1989[2]). Consequently, the coexistence of material progress and sustainability is no longer self evident: this explains the resistance which the idea of sustainability faces in western societies (Toman, Pezzey, Krautkraemer 1994).

What would the implications be of extending the notion of primary social assets to natural resources[3] and the right to be employed, while considering intergenerational and intragenerational equity? At any given time, within any particular society, two series of circumstances are rare: pooled resources of physical or stored expertise and opportunities to increase these resources, that is to say, employment in which man's efforts can be put to use, taking socio-technical organization into account. Equity requires these resources to be shared out equally, while making sure that access to these natural resources will be guaranteed for future generations. If, at any given time, all energy supplies and opportunities for pooled human resources (jobs) were unavailable to even a fraction of the population, then, a further sharing out would be necessary. Take, for example, the case of employment: the sharing out of available jobs between all members of society likely to fill them also implies the sharing out of corresponding primary incomes for these jobs; in other terms, the reduction in inequalities of these incomes. This proposition clashes with one of universal subsidy (Van Parijs 1991; Bresson 1993; Ferry 1995) which acts as a counterbalance to the impossibility or lack of willingness to redeploy energy supplies or opportunities for pooled resources. It is a matter of asserting the priority of the right to work which corresponds to a *primary social asset* in

---

could be more of an obstacle, at the best, a meaningless distraction, at the worst, a temptation to accept facility and emptiness. (...) Unlike the principle of equality, intertemporal preference has not intrinsic ethical interest. It is introduced in a purely *ad hoc* manner to lighten the consequences of the utility criterion. " (Rawls 1987, p. 331-337, my translation).

[2] Norton differentiates four possible strategies concerning natural resources: exploitationism, conservationism, and two variants of preservationism: naturalist and extensionist. This classification partially tallies with the distinction between the anthropocentric and ecocentric approaches of sustainability (Hatem1990).

[3] Certains authors have emphasized that natural resources are conditions for the acquisition of primary social assets (Pearce 1987) or that equality should be considered in terms of 'basic capabilities' (Sen 1993).

constrast with the *palliative of its non-compliance*[4]. While universal subsidy aims to create a new citizenship because the former one is not respected, it is our opinion that a trend towards equity arises from relationships within a solidarity that is constantly and simultaneously redeploying jobs and income.

Since the environmental crisis and the social crisis are combined, how could a renunciation of economic growth be initiated in industrialized countries, which are top of the list of those responsible for environmental damage, from the very moment when essential needs were being satisfied, and how could the demands of social justice be met, now and over the years to come? A strategy such as this comes up against capital interests to which economic growth supplies constant opportunities to put itself to the fore and for which concern for the environment simply represents a timely market development. In addition, a strategy such as this implies a deconstruction of collective make believe where 'better living' is made out to mean 'having more' (Passet 1985, p. 833). Lastly, a strategy such as this is the opposite extreme of the most widely held ideas about sustainability. Nevertheless, "the expression 'sustainable development' is correct if applied to economy, that is to say a *physically stable qualitative improvement* maintained under stable conditions defined by the limits of ecosystems" (Daly 1992, p. 10).

A long term environmentally and socially sustainable development strategy is only meaningful as long as the primary use of the productivity gains is not to generate an everlasting production and consumption growth, but to reduce working hours for everyone in such a way as to give opportunities of employment to all those wishing to find work and to give everyone the chance of experiencing other forms of wellbeing besides eternal consumption growth. Concepts of equity and solidarity both between and within generations should thus be harmonised.

In the short term, in order to rapidly solve the unemployment problem in industrialized countries, while considering the need to deliberately give up high growth because of the risk of draining nonrenewable resources and the responsibility of planning a fairer distribution between all the planet's inhabitants, we have made a study of the funding of job creation through the reduction in income inequalities.

In order to give our approach a theoretical basis, we propose a simple mathematical model for the reduction in the hierarchical income scale which would make a decrease in unemployment possible while protecting ecological balance from the threats of an excessively high economic growth. This model can function in two ways: for a given unemployment rate, it allows us either to determine the scale of the reduction in inequalities to be achieved in order to cut

[4] This point is developed in Harribey 1996-a and 1996-b.

down unemployment within a given period, or to determine the time necessary to reduce it pending a decision about socially desirable and/or sustainable income hierarchy.

We test our model on the French economy by following three stages. First, we apply it to existing distribution of household incomes and then on wage distribution alone. In both cases, it turns out that the elimination of unemployment is a distributional problem and not one due to wealth level. Finally, we perform a third simulation to reduce the consequences in terms of consumption growth of an excessively high rise in low and average household incomes whose marginal propensity to consume is higher than that of people whose incomes are penalized by a reduction of inequalities: in this case, the period of time necessary to cut down unemployment is reduced and improvement in the standard of living of the lower classes depends to a small extent on raising their purchasing power and to a larger extent on a reduction in working hours.

13.1 The model for a reduction in income inequalities with the aim of financing a reduction of working hours

Given that N is the active working population divided into n income classes; as the classes contain the same number of employees, then the proportion of these employees in each income class is: $q = 1/n$. The total of incomes paid to households is W. The hierarchy between the average income of each income bracket is expressed by the relative differences: $a_1, a_2, a_3, ..., a_n$, with the conditions $a_1 = 1 < a_2 < a_3 < ... < a_n$. The proportion of unemployed in each class[5] in relation to the total number of unemployed is: $h_1, h_2, h_3, ..., h_n$, with $h_1 + h_2 + h_3, ..., + h_n = 1$. The unemployment rate in relation to the active working population is h^*. If the active working population increases by a rate $h = h^*$, unemployment disappears. Take the case of a restricted hierarchy defined thus: $a'_1, a'_2, a'_3, ..., a'_n$, with $a'_1 = 1 < a'_2 < a'_3 < ... < a'_n$, and $a'_1 \leq a_1$, $a'_2 \leq a_2, ..., a'_n < a_n$.[6] Average incomes of the first income class before and after restriction of income hierarchy are respectively called R and R'.

---

[5] Assuming that if these unemployed people found a job, they would be split into the income classes in proportion $h_1, h_2, h_3, ..., h_n$.

[6]. At least one of the hierarchical differences in the new distribution must be strictly inferior to the corresponding hierarchical differences in the former distribution for there to be any restriction of the income hierarchy.

We first consider that the production and the total of all distributed incomes have not varied between the two situations. So, before restriction of the income hierarchy:

$$W = q\ N\ R + a_2\ q\ N\ R + a_3\ q\ N\ R + \ldots + a_n\ q\ N\ R$$

$$R = \frac{W}{q(1+a_2+a_3+\ldots+a_n)N} = \frac{n}{N}\frac{W}{(1+a_2+a_3+\ldots+a_n)} = \frac{W}{qN\sum_{i=1}^{n} a_i} = \frac{n}{N}\frac{W}{\sum_{i=1}^{n} a_i}.$$

$$R = \frac{W}{q(1+a_2+a_3+\ldots+a_n)N} = \frac{n}{N}\frac{W}{(1+a_2+a_3+\ldots+a_n)} = \frac{W}{qN\sum_{i=1}^{n} a_i} = \frac{n}{N}\frac{W}{\sum_{i=1}^{n} a_i}. \quad (13.1)$$

After restriction of the income hierarchy:

$$W = [q\ (1 + a'_2 + a'_3 + \ldots + a'_n) + h\ (h_1 + a'_2\ h_2 + a'_3\ h_3 + \ldots + a'_n\ h_n)]\ N\ R'$$

$$R' = \frac{W}{[q(1 + a'_2 + a'_3 + \ldots + a'_n) + h(h_1 + a'_2 h_2 + a'_3 h_3 + \ldots + a'_n h_n)]N}$$

$$= \frac{W}{\left[\left(q\sum_{i=1}^{n} a'_i\right) + \left(h\sum_{i=1}^{n} h_i a'_i\right)\right]N} \quad (13.2)$$

The variation of the basic average income in the first income class is expressed thus:[7] (13.3)

$$\frac{R'}{R}=\frac{q(1+a_2+a_3+...+a_n)}{q(1+a'_2+a'_3+...+a'_n)+h(h_1+a'_2h_2+a'_3h_3+...+a'_nh_n)}=\frac{q\sum a_i}{q\sum a'_i+h\sum h_i a'_i}.$$

We shall set down this expression

$$\frac{R'}{R}=1+r$$

and call $1+r$ the *multiplying coefficient of solidarity*. By calling a'_2/a_2 , ..., a'_n/a_n , the coefficients of the reduction in inequalities applied to each of the n - 1 last income classes,

- the variation of the average income in the 2nd income class is:

$$\frac{a'_2 R'}{a_2 R}=\frac{a'_2}{a_2}(1+r);$$

- the variation of the average income in the nth income class is:

$$\frac{a'_n R'}{a_n R}=\frac{a'_n}{a_n}(1+r);$$

We thus obtain the n multiplying coefficients of the average income of each class and are then able to extract the income necessary for payment to the hN newly employed active workers. The variation rates are immediately deducted. If the output, the total of distributed incomes and the hourly work productivity do not vary then the per capita working hours will vary on average:

$$1+t=\frac{1}{1+h}$$

---

[7] Remark: it is not indispensable for the population to be divided into equal fractions; if they were different such as: $q_1, q_2, q_3,..., q_n$, then the coefficient of solidarity would be expressed as follows:

$$1+r=\frac{q_1+q_2a_2+q_3a_3+...+q_na_n}{q_1+q_2a'_2+q_3a'_3+...+q_na'_n+h(h_1+a'_2h_2+a'_3h_3+...+a'_nh_n)}=\frac{\sum q_i a_i}{\sum q_i a'_i+h\sum h_i a'_i}.$$

We now suppose that, between the two situations, before and after restriction of the income hierarchy, output had been multiplied by 1+y. We put forward the following hypotheses:

- the sharing out of distributed household incomes and social surplus does not change;
- the wages and other individual incomes all vary at an equal rate z; irrespective of the variation of incomes linked to the reduction in inequalities;
- hourly productivity varies at a rate x;
- per capita working hours vary at a rate t;
- the number of jobs varies at a rate h.

We would either have: $(1 + y) = (1 + x)(1 + t)(1 + h) = (1 + z)(1 + h)$, or $(1 + z) = (1 + x)(1 + t) = (1 + y)/(1 + h)$. Average income in the 1st income class as shown above could therefore vary thus: (13.4)

$$\frac{R'}{R} = (1+r)(1+z) = \frac{(1+r)(1+y)}{1+h};$$

in the 2nd income class: (13.5)

$$\frac{a'_2 R'}{a_2 R} = \frac{a'_2}{a_2}(1+r)(1+z) = \frac{a'_2}{a_2}\frac{(1+r)(1+y)}{1+h};$$

in the nth income class: (13.6)

$$\frac{a'_2 R'}{a_2 R} = \frac{a'_2}{a_2}(1+r)(1+z) = \frac{a'_2}{a_2}\frac{(1+r)(1+y)}{1+h};$$

In each income class the average income can only increase if the product of the multiplying coefficient of solidarity and the multiplying coefficient of individual incomes is higher than the inverse of the coefficient of reduction in inequalities in the income class i thus shown; $(1 + r)(1 + z) > a_i / a'_i$. Individual working hours vary on average by: (13.7)

$$1+t = \frac{1+y}{(1+x)(1+h)} = \frac{1+z}{1+h}.$$

13.2 Application of the model

13.2.1 Application of the model as applied to all incomes

We apply the model of existing household incomes distribution of which the hierarchy ranges from 1 to more than 17 between the outer decile with an exponential widening of the gaps from the 5th decile onwards (Table 13.1): we simulate the impact that a restriction of the income hierarchy would have on jobs thus making this hierarchy an extension of the trend to have wider gaps for the first four deciles (Table 13.2). The prevalent trend in the first four deciles is estimated by the relation $y = 1.84335 \text{ Ln } x + 0.95712$ (x being the deciles and y the hierarchical differences, with a coefficient of linear correlation 0.998). The restricting process is then spread over four years at a stable rate.

Table 13.1 The taxable income in France in 1990.

1 Deciles	2 Highest limit of the class (in francs)	3 Average taxable income (in francs)	4 Income hierarchy	5 Cumulation of distributed income (in %)
		24 900	1	
D1	43 141			1.68
		54 200	2.18	
D2	63 773			5.33
		72 600	2.92	
D3	81 143			10.22
		89 500	3.59	
D4	98 277			16.25
		107 800	4.33	
D5	117 336			23.51
		128 000	5.14	
D6	139 109			32.13
		152 300	6.12	
D7	165 614			42.39
		182 900	7.35	
D8	203 013			54.71
		230 800	9.27	
D9	267 230			70.26
		301 500	17.72	
95th centile	347 953			80.42
		581 000		
				100
Together		148 500		

Source: INSEE, 1995-a, extract of table 2p. 14, except the 4th and 5th columns which we calculated.

The unemployed represent $h^* = 3\ 114.7/21\ 798 = 14.29\%$ of the active working population. Starting from the division of registered job seekers into socio-professional categories (Table 13.3), we shall assume that each job seeker would find a job in his category with the exception of former farmers, craftsmen, commercial traders and business managers, and lastly those who have never been employed (Table 13.4)[8].

Table 13.2 Timing for the reduction in income inequalities

Décils	Existing real hierarchy	Progressively restricted hierarchy			
		1st year	2nd year	3rd year	4th year
D1	1	1	1	1	1
D2	2.18	2.18	2.18	2.18	2.18
D3	2.92	2.92	2.92	2.92	2.92
D4	3.59	3.59	3.59	3.59	3.59
D5	4.33	4.27	4.21	4.16	4.10
D6	5.14	4.96	4.78	4.61	4.45
D7	6.12	5.74	5.39	5.06	4.75
D8	7.35	6.68	6.06	5.51	5.00
D9	9.27	8.03	6.96	6.03	5.23
D10	17.72	13.18	9.81	7.30	5.43

Table 13.3 Unemployment by CSP (march 1994).

CSP	Effectives (in thousands)
Farmers	3.5
Craftsmen, com. traders, busin. dir.	66.0
Exec. and higher intell. professions	159.5
Intermediary professions	363.8
Employees	1 094.1
Workers	1 098.9
Those who have never had a job	328.9
Total	3 114.7

Source: INSEE, TEF, 1995-96, p. 73.

[8] We divide the job seekers into income classes: – workmen and employees: 10% in the 2nd household decile, 40% in the 3rd, 40% in the 4th and 10% in the 5th; – intermediary professions: 10% in the 4th decile, 40% in the 5th, 40% in the 6th and 10% in the 7th; – senior executives and higher intellect professions: 10% in the 6th decile, 20% in the 7th, 32% in the 8th and 38% in the 9th; – farmers, craftsmen, commercial traders and business directors: 20% in the 2nd decile, 50% in the 3rd, 20% in the 4th and 10% in the 5th; – job seekers having never worked: 10% in the 2nd decile, 20% in the 3rd, 20% in the 4th, 20% in the 5th, 15% in the 6th, 10% in the 7th, 4.5% in the 8th and 0.5% in the 9th.

Let us suppose that this plan were to be put into effect over a period of four years. During each of the first three years, one third of the existing unemployed would find a job, and during the fourth year, an equivalent number (corresponding to a fourth third) would become newly registered job seekers who would find jobs. To say that the cutting down of unemployment is equal to the creating of jobs for four thirds of the existing number of unemployment comes down to supposing that the number of unemployment never goes down in proportion to the number of jobs created. The multiplying coefficient 1.33 adopted is higher than that retained by the General Plan Commission[9] in order to make allowances for the fact that, in reality, the reduction in working hours does not mean a proportionally higher number of jobs since the absolute value of employment elasticity in relation to the number of hours is often lower than the unit.

Table 13.4 Integration of the unemployed into the income receiving population.

Décils	Farmers, Craftsmen, Comm. traders, business,. directors	Executives and higher intellect professions	Intermediary professions	Workers, employees	Unemploy. never had a job	Total	Proportion
D1							
D2	13 900			219 300	32 890	266 090	0.08543
D3	34 750			877 200	65 780	977 730	0.31391
D4	13 900		36 380	877 200	65 780	993 260	0.31889
D5	6 950		145 520	219 300	65 780	437 550	0.14048
D6		15 950	145 520		49 335	210 805	0.06768
D7		31 900	36 380		32 980	101 170	0.03248
D8		51 040			14 800	65 840	0.02114
D9		60 610			1 645	62 255	0.01999
D10							
Total	69 500	159 500	363 800	2 193 000	328 900	3 114 700	1

[9] The General Plan Commission indicates that if working hours had continued to drop after 1986 at the rate of 0,8% a year as from 1970 to 1986, and in view of productivity gains, 312 000 jobs would have had to be created which would have reduced unemployment by 250 000; the coefficient is therefore 1,25 (Commissariat General du Plan, 1993, p. 268).

Table 13.5 Recapitulation of the simulation of the reduction in income inequalities over 4 years.

1 Classes	2 Multiplying coefficient of average income	3 Variation rate of average income	4 Multiplying coefficient of working hours	5 Variation rate of working hours	6 Multiplying coefficient of income in units of work (2)/(4)	7 Variation rate of income in units of work
1	1.3967	+ 39.67%	0.961543	- 3.85%	1.45256	+ 45.26%
2	1.3967	+ 39.67%	0.827139	- 17.29%	1.68859	+ 68.86%
3	1.3967	+ 39.67%	0.601745	- 39.83%	2.32108	+ 132.11%
4	1.3967	+ 39.67%	0.597976	- 40.20%	2.33571	+ 133.57%
5	1.3225	+ 32.25%	0.758521	- 24.15%	1.74353	+ 74.35%
6	1.2062	+ 20.62%	0.851709	- 14.83%	1.41621	+ 41.62%
7	1.0837	+ 8.37%	0.905474	- 9.45%	1.19683	+ 19.68%
8	0.9501	- 4.99%	0.924272	- 7.57%	1.02794	+ 2.79%
9	0.7880	- 21.20%	0.926283	- 7.37%	0.85071	- 14.93%
10	0.4280	- 57.20%	0.961543	- 3.85%	0.44512	- 55.49%

Note 1. In so far as each class of the population has assimilated an unequal number of unemployed, the population is no longer divided into deciles but as follows for the 10 classes respectively: 8.4%; 9.8%; 13.4%; 13.5%; 10.6%; 9.5%; 8.9%; 8.7%; 8.7%; 8.4%.

We have chosen not to be hindered by a coefficient of specific elasticity coming into play in the matter of softening the consequences of reduced working hours because the main obstacle for a reduction in the latter, in firms, is put forward in our simulation which is developed by distributing a constant overall income, the only possible increase arising from a prospective economic growth.

Precisely, we can suppose that the spreading out of the abolition of unemployment and the reduction in inequalities over a four years period would bring about an average annual growth rate of hourly productivity of 2.5%. In order to soften the social impact brought about by the lowering of income in the penalized social groups by the reduction in inequalities, economic growth is supposed to be used to raise individual incomes at a uniform annual rate of 1.5%, the sharing out of *distributed income/social surplus* will remain unchanged.

At the end of four years the active working population receiving income will have increased to 4 152 933 or 19.05%. The average reduction in per capita working hours over four years would be: $1.015^4 / 1.1905 \, . \, 1.025^4 - 1 = -19.23\%$.

The reduction in working hours is not equal for everybody. The reason for this is that the structure of the unemployed population does not correspond to the structure of the active working population, the less qualified categories being over represented. From then on, the reduction in working hours for those already employed (which would be necessary for the integration of the registered job seekers) should be as intense as the number of the unemployed in the corresponding categories is high. When all the job seekers have found satisfaction, the reduction in working hours could be carried out in an equal manner between all the groups of workers progressing alongside productivity gains.

13.2.2 Application of the model for wages

The application of the model for wage-earned income alone does not aim at abandoning the objective of linking all incomes to the funding of job creation (made possible by the reduction in working hours) but to put an end to an ambiguity which weighs heavily on debates concerning this latter question: for many, the reduction in working hours would not be possible without imposing a reduction in wages for most wage-earners, sparing not even those of the most modest means. We refute this assertion and shall show that discussions on compensation or non compensation of wages stemming from the reduction in working hours can easily be clarified.

We start with the distribution of wages after tax corresponding to the full-time wage-earning workforce in the private, semi-public and public sectors. By applying the same method as for all incomes, we shall seek to measure the impact of a reduction in wage inequalities on employing the unemployed. We shall use the same hypotheses as before, knowing that productivity increases, on average, by 2.5% a year and that an annual economic growth of 1.5% helps to raise all wages at a uniform rate, while a third of the unemployed will find jobs each year; lastly, we can presume that the creation of jobs will incite an influx of additional job seekers equivalent to a fourth third finding work during the fourth year.

The progression of hierarchical differences between income classes (Table 13.6) is almost perfectly linear up to the fifth wage class (adjustment: $y = 0.1359 x + 0.8579$, with a coefficient of correlation of 0.99972). Above that, the progression takes on an exponential shape. For this reason, we shall test the impact that a restriction of the wage hierarchy would have on employment so that this hierarchy would be brought down to the level it would have if the progression of hierarchical differences continued at a linear rate beyond the 5th wage class (Table 13.7). Jobs seekers would then be integrated into the different income classes[10] (Table13.8).

[10] We divide the job seekers as follows: – workmen and employees: 15% in the 1st class, 20% in the 2nd, 30% in the 3rd, 20% in the 4th, 10% in the 5th and 5% in the 6th; – intermediary professions: 5% in the 3rd class, 15% in the 4th, 25% in the 5th, 30% in the 6th, 15% in the 7th and 10% in the

Table 13.6 Distribution of net wages in 1992

Wage classes	Distribution of wage-earners	Upper limit of the class (in francs)	Average wage of the class (in francs)	Hierarchy of average wages
			60 000	1
1	0.0851	62 900		
			67 300	1.122
2	0.0886	71 700		
			75 900	1.265
3	0.0906	80 100		
			84 050	1.401
4	0.0937	88 000		
			92 400	1.540
5	0.0979	96 800		
			102 000	1.700
6	0.1022	107 200		
			114 250	1.904
7	0.1074	121 300		
			132 600	2.210
8	0.1185	143 900		
			167 650	2.794
9	0.1141	191 400		
			278 000	4.633
10	0.1019			
	1			

Sources: For the information on wages in the private and semi-public sectors: INSEE, TEF, 1995-96, p. 83; for those on wages in the public sector: Quarre D. 1995, p. 245. We have grouped together information concerning all types of wages in this table (for the methodology, cf. Harribey 1996-b).

At the end of the four years, the simulation of a reduction in wage inequalities shows (Table 13.9) a rise in wages for 70% of wage-earners and a rise in the hourly wage for more than 80% of wage-earners, as unemployment is relieved by the reduction in working hours. Thus, the only professions who would be penalized by a sharp reduction in wage inequalities would be the higher paid among senior executives and intellect professions. Only about 10% or less of wage-earners situated immediately below the better paid senior executives on the wage scale would have a reduction in their working hours which would not be

8th; – senior executives and higher intellect professions: 20% in the 6th class, 25% in the 7th, 25% in the 8th, 20% in the 9th and 10% in the 10th; – farmers, craftsmen, commercial traders and business directors: 10% in the 1st class, 20% in the 2nd, 30% in the 3rd, 25% in the 4th, 10% in the 5th et 5% dans la 6°; – job seekers having never worked: 10% in the 1st class, 15% in the 2nd, 20% in the 3rd, 15% in the 4th, 12% in the 5th, 10% in the 6th, 8% in the 7th, 5% in the 8th, 3.5% in the 9th and 1.5% in the 10th.

compensated by a rise in the hourly wage or, more explicitly, would suffer a drop in wages which would be more than compensated by a reduction in their working hours. Wage inequalities and, even more blatantly, income inequalities are so great in France that their reduction would safeguard the maintenance and even the progression of purchasing power for the great majority of wage-earners as working hours are reduced.

Table 13.7 Restriction of the wage hierarchy.

Classes	Existing hierarchy	Restricted hierarchy at the end of the 1st year	Restricted hierarchy at the end of the 2nd year	Restricted hierarchy at the end of the 3rd year	Restricted hierarchy at the end of the 4th year
1	1	1	1	1	1
2	1.122	1.122	1.122	1.122	1.122
3	1.265	1.265	1.265	1.265	1.265
4	1.401	1.401	1.401	1.401	1.401
5	1.540	1.540	1.540	1.540	1.540
6	1.700	1.6932	1.6864	1.6797	1.673
7	1.904	1.8798	1.8559	1.8323	1.809
8	2.210	2.1405	2.0733	2.0081	1.945
9	2.794	2.5956	2.4113	2.2401	2.081
10	4.633	3.8533	3.2049	2.6656	2.217

Note : The last column (classes 6 to 10) is obtained by applying the relation $y = 0.1359\,x + 0.8579$. Starting from the 6th class, the fourth roots of the divisions of the last and the 2nd columns give the annual multiplying coefficient of the new hierarchical levels (3rd column); for example for the 10th class: $(2.217/4.633)^{1/4} = 0.8317$ which we multiply by 4.633 to obtain 3.8533, etc.

The putting into practice of a sharp reduction in income inequalities would lead to a rapid and substantial rise in lower incomes and this, all the more so, since the bulk of incomes would be used and not only wages; if the principle of solidarity is applied only to wages, the scope of the rise in lower and average wages would be very small, but if the principle is applied to all incomes, the scope would be much greater. Thus, the marginal propensity to consume of those with modest incomes is greater than those whose incomes would be cut down by the reduction in inequalities. It is therefore very probable that the increase in low incomes would be followed by a significant increase in consumation and that, by means of a phenomenon of multiplication, a much greater economic growth than has been seen, on average, for the last two decades would be triggered off. The rapid cutting down of unemployment, the rise in lower and average wages and the return of economic growth could therefore run the risk of being in contention with the search for environmental sustainability. Is it possible to go beyond this contradiction?

Table 13.8 Integration of the unemployed into the active wage-earning population.

Classes	Farmers, Craftsmen, Comm. traders, busin. directors	Executives and higher intellect professions	Intermediary professions	Workers, employees	Unemploy. never had a job	Total	Proportion
1	6 950			328 950	32 890	368 790	0.11840
2	13 900			438 600	49 335	501 835	0.16112
3	20 850		18 190	657 900	65 780	762 720	0.24488
4	17 375		54 570	438 600	49 335	559 880	0.17975
5	6 950		90 950	219 300	39 468	356 668	0.11451
6	3 475	31 900	109 140	109 650	32 890	287 055	0.09216
7		39 875	54 570		26 312	120 757	0.03877
8		39 875	36 380		16 445	92 700	0.02976
9		31 900			11 511	43 411	0.01394
10		15 950			4 934	20 884	0.00670
Total	69 500	159 500	363 800	2 193 000	328 900	3 114 700	1

Table 13.9 Recapitulation of the reduction in wages inequalities simulation in 4 years

1 Classes	2 Multiplying coefficient of average wage	3 Variation rate of average wage	4 Multiplying coefficient of working hours	5 Variation rate of working hours	6 Multiplying coefficient of unitary wages (2)/(4)	7 Variation rate of unitary wages
1	1.0567	+ 5.67%	0.69224	- 30.78%	1.52650	+ 52.65%
2	1.0567	+ 5.67%	0.63743	- 36.26%	1.65775	+ 65.78%
3	1.0567	+ 5.67%	0.54765	- 45.24%	1.92952	+ 92.95%
4	1.0567	+ 5.67%	0.62583	- 37.42%	1.68847	+ 68.85%
5	1.0567	+ 5.67%	0.72457	- 27.54%	1.45838	+ 45.84%
6	1.0400	+ 4.00%	0.76791	- 23.21%	1.35431	+ 35.43%
7	1.0040	+ 0.40%	0.87338	- 12.66%	1.14955	+ 14.96%
8	0.9300	- 7.00%	0.89845	- 10.16%	1.03512	+ 3.51%
9	0.7871	- 21.29%	0.92980	- 7.02%	0.84653	- 15.35%
10	0.5057	- 49.43%	0.94418	- 5.58%	0.53560	- 46.44%

13.2.3 Towards a sustainable model

We test our model for the last time, hoping to bring together the maximum number of guarantees in relation to environmental sustainability while re-establishing conditions of social solidarity:

- an intermediary reduction in inequalities situated between the maintenance of the status quo and the reduction corresponding to the first two tests of the above model;
- a progression in the taxation on higher incomes and capital which would be all the more substantial as the reduction in direct income inequalities would be small;
- a progression in the taxation on low and average incomes which would be all the smaller as the reduction in direct income inequalities would be small and
- at the least, a preservation of purchasing power for all low and average incomes and, at the most, a progression of purchasing power all the smaller as the reduction in working hours would be large; this comes down to dividing out the bulk of income made disposable by the reduction in inequalities on a number of unemployed people who have been integrated into the wider working population, or in other words, an equivalent to the shortening of the integration period for the unemployed.

By referring to the freezing of wages and purchasing power which began in France in 1982, the GDP increased by 27.56% from the beginning of 1982 to the end of 1994 while the purchasing power of the 'SMIC' increased by only 19.9%.[11] By setting as a deadline the increase of the average income of the lowest income class for the two year period for which we can now test our model for the reduction in inequalities, this deficit (accumulated by the increase in purchasing power) versus to the progression of the GDP (the necessary catching up), is given by the division progression of GDP/progression of SMIC: 1.27564/1.16898 = 1.09124. Presuming that a moderate economic growth insures an additional average annual increase of 1.5%, the global progression of the lowest average income is: $(1 + r)(1 + z) = 1.09124 \,.\, 1.015^2 = 1.12422$, or slightly more than 6% during each of the two years (+ 6.029% a year).

We shall test the model while insuring a decreasing progression of average incomes. The latter be decrease by one percentage point for each decile from the first to the seventh: 2: 5%; 3: 4%; 4: 3%; 5: 2%; 6: 1%; 7: 0%. For the last three deciles the reduction in average incomes will be such that they will reach the level corresponding to the restriction of the income hierarchy already applied in the column for the 2nd year of table 13.2. The choice of only cutting down

[11] Sources: INSEE, 1987, p. 93, 1993-1994, p. 89, 1995-1996, p. 101, for the GNP; INSEE, 1995-b, table 03.06, p. 59, for the productivity; INSEE, 1995-1996, p. 87, for the progression of purchasing power of the gross 'SMIC' (minimum hourly wage in France).

average incomes beyond the 7th decile is aimed at not penalizing more households than in the preceding estimations. The results arrived at are summed up in table 13.10.

It appears then that it is possible to reduce the deadline for cutting down unemployment to two years provided that the rise in lower and average incomes, made possible by the reduction in inequalities, is limited.

- The multiplying coefficient of solidarity shows that the bulk of income made disposable by the reduction in inequalities does not entail a rise (at the least, equal) in basic income as was the case in the preceding simulations; in fact, this multiplying coefficient $(1 + r)$ is higher than that of the basic income even though the latter includes the rise which stems directly from economic growth: $1 + r = 5.961/5.28888 = 1.12708$; $r_2/r = 1.12422 = (1 + r)(1 + z)$,

  and $1 + z = 1.12422/1.12708 = 0.99746$.

In other words, the increase in basic income of 12.71% which would result from applying only the multiplying coefficient is diminished by about 0.25%. This drop can be analysed as the combination of a rise resulting from economic growth gains and from a more significant rise in taxes; if y is the economic growth rate and f the growth rate of tax burden: $(1 + z)(1 + f) = 1 + y$, and $1 + f = 1.015^2/0.99746 = 1.032847$.

- The average reduction in working hours over a two year period is: $t = 1.015^2/1.190533.\ 1.025^2 - 1 = -17.63\%$.

Conclusions

We have made our model work for a new sharing out of employment and income by adopting certain values for the economic growth rate, productivity growth and the deadline for the cutting down of unemployment. How can we justify these hypotheses?

The annual rate of production growth of 1.5% corresponds on average to the downward trend of economic growth experienced by most western countries in the last two decades. This rate is thus considered by most observers to be the *intolerable minimum threshold* above which one must rise in order to relive the 'Trente Glorieuses' (Fourastie 1985) through 'Vingt Merveilleuses' (Izraelewicz 1994). We have chosen it deliberately as the tolerable maximum threshold because it is amply sufficient for covering the needs arising from the

Table 13.10 The sharing out of incomes and jobs over two years.

1 Population Recieving Incomes divided into deciles at the beginning of 1st year	2 Hierarchy of average incomes at the beginning of 1st year	3 Total distributed income within each decile (1).(2) R	4 Division of unemployed in Proportion to the total during the 2 years	5 Population receiving incomes after the employment of the unemployed (1) + (4) hN	6 Multiplying coefficient of average incomes over a 2 year period for the three last (7). 1,12422/(2)	7 Restricted Income hierarchy at the end of the 2nd year (2).(6)/1, 12422 except for the three last	8 New Distributed Income (5).(7) R_2	9 New average income (7) R_2 = (7)(1+r)(1+z) R	10 Variation Rate of Working hours[12]	11 Variation rate of income by unit of work[13]
0.1 N	1	0.1 NR	-	0.1 N	1.12422	1	0.1 NR2	1.12422R	- 1.94%	+ 14.65%
0.1 N	2.18	0.218 NR	0.113907	0.116277 N	1.10250	2.14	0.24883 NR2	2.40583R	- 15.67%	+ 30.73%
0.1 N	2.92	0.292 NR	0.418544	0.159810 N	1.08160	2.81	0.44907 NR2	3.15906R	- 38.64%	+ 76.27%
0.1 N	3.59	0.359 NR	0.425192	0.160760 N	1.06090	3.39	0.54498 NR2	3.81111R	- 39.00%	+ 73.93%
0.1 N	4.33	0.433 NR	0.187305	0.126766 N	1.04040	4.01	0.50833 NR2	4.50812R	- 22.65%	+ 34.50%
0.1 N	5.14	0.514 NR	0.090241	0.112895 N	1.02010	4.66	0.52609 NR2	5.23887R	- 13.14%	+ 17.44%
0.1 N	6.12	0.612 NR	0.043309	0.106189 N	1.00000	5.44	0.57767 NR2	6.11576R	- 7.66%	+ 8.39%
0.1 N	7.35	0.735 NR	0.028185	0.104028 N	0.92691	6.06	0.63041 NR2	6.81277R	- 5.74%	- 1.67%
0.1 N	9.27	0.927 NR	0.026650	0.103808 N	0.84407	6.96	0.72250 NR2	7.82457R	- 5.54%	- 10.64%
0.1 N	17.72	1.772 NR	-	0.1 N	0.62238	9.81	0.981 NR2	11.0286R	- 1.94%	- 36.53%
1 N	59.62	5.962 NR	1.333333	1.190533 N		46.28	5.28888 NR2			

[12] The multiplying coefficients of working hours are calculated as follows: $1.015^2/[(5^{th}$ column of table$/1^{st}$ column of table$)\ 1.025^2]$.

[13] The multiplying coefficients of income by unit of work are calculated as follows: multiplying coefficients of average incomes/multiplying coefficients of working hours.

demographic growth in industrialized countries and secondly because it has a much greater probability of long-term achievement than would a higher rate.[14]

The hourly productivity growth of 2.5% that we set down corresponds to the existing trend in Europe. Its lower level in opposition to the post-war period must be linked to the slightest economic growth. The maintenance of moderate economic and productivity gains are therefore, in our opinion, hypothetically linked.[15]

In the different simulations which we have carried out, we adopted a deadline of two to four years for cutting down unemployment in France. Is this a reasonable deadline? Could our conclusion be discredited by its briefness? We notice that it falls into line with those that are always heralded (but never respected) by those in charge of economic and political matters.[16] We deemed it necessary to test its realism by means of an alternative approach to those already tried. Moreover, this deadline corresponds to that adopted by parallel simulations carried out elsewhere and confirmed by our own studies.[17]

The model presented here simply aims to show that the funding of the job creation necessary for a substantial cutting down of unemployment is possible even in the absence of sharp economic growth. Problems arising from economic development, those from exclusion as for those arising from the damage done to ecosystems can not be solved within this form of development. Sustainability is only meaningful if it is conceived as a distribution issue: distribution of limited natural resources between generations, distribution of work and produced wealth within the present generation. Instead of living in the throes of a moderate economic growth, it is possible to profit from this situation and trigger off social and cultural changes in the ways we comprehend the improvement of well-being. Thus, the principle of *responsibility* towards life, the principle of *solidarity* between individuals of the same generation and with those of future generations, and the principle of *sparing* limited resources and manpower, spell out sustainability ethics. If these principles were to be pushed aside, we should fear a

[14] We refer to Maddison's most recent study which indicates an average annual growth rate of GDP of Western Europe of 2.2% for the period from 1820 to 1992 just as from 1973 to 1992, and a per capita growth rate of GDP of 1.5% from 1820 to 1992 and of 1.8% from 1973 to 1992. The average growth rate of GDP for France from 1973 to 1992 is 1.7% a year, that of a cross section of 12 western European countries is 1.8%. (Maddison 1995, p. 62 and 64).

[15] Maddison provides the following average annual growth rate of hourly productivity for the period from 1973 to 1992: 2.7% in France and 2.3% in 12 western European countries. (Maddison 1995, p. 80).

[16] To take an example outside France, in the course of negotiations for a new 'employment pact' which took place in January 1996 between German management and Trade Unions, the aim was to reduce unemployment by half (then 4 million in number) by the year 2000.

[17] See especially Nikonoff 1995.

situation in which durable development would only be a durable growth, 'a provoking slogan' (Daly 1992, p. 11) and thus simply a 'concept alibi' (Latouche 1994). On the other hand, if they were adopted, a serious research programme should be set up concerning the notion of needs that economic theory has never be able to characterize otherwise than by the simplistic statement " they are (almost by nature) unlimited " which justifies in advance the pursuit of an endless economic growth. Finally, the obstacles along the path of sustainability and justice are less economic than social, political and cultural.

References

Bidet J. (1995), John Rawls et la théorie de la justice, Paris, PUF, Actuel Marx Confrontation.

Bresson Y. (1993), L'après-salariat, Une nouvelle approche de l'économie, Paris, Economica, 2° éd.

Commissariat Général du Plan (1993), L'économie française en perspective, Paris, La Découverte, La Documentation française.

Daly HE. (1990), 'Toward some operational principles of sustainable development', Ecological Economics, vol. 2, n° 1, avril, pp. 1-6. (1992),

'Il n'y a pas de croissance durable', Transversales Science/Culture, n° 13, janvier-février.

Dupuy JP. (1992), Le sacrifice et l'envie, Le libéralisme aux prises avec la justice sociale, Paris, Calmann-Lévy.

F.A.O. (1993), World Watch List for Animal Diversity, Edited by Ronan Loftus and Beate Scherf, 1st edition, Rome, november.

Ferry JM. (1995), L'allocation universelle, Pour un revenu de citoyenneté, Paris, Ed. du Cerf.

Fourastie J. (1985), Les Trente Glorieuses, Paris, Hachette.

Georgescu-Roegen N. (1995), La décroissance, Entropie-Ecologie-Economie, Paris, Ed. Sang de la terre.

Gorz A. (1988), Métamorphoses du travail, Quête du sens, Critique de la raison économique, Paris, Galilée.

Harribey JM. (1996-a), 'Théorie de la justice, revenu et citoyenneté', La Revue du M.A.U.S.S. semestrielle, n° 7, 1er semestre, pp. 188-198.

Harribey JM. (1996-b), Développement soutenable et réduction du temps de travail, Analyse critique appliquée au cas de la France, Thèse de Doctorat en Sciences Economiques, Université de Paris I-Panthéon-Sorbonne.

Hatem F. (1990), 'Le concept de "développement soutenable"', Economie prospective internationale, n° 44, 4° trimestre, pp. 101-117.

INSEE (1987, 1993-1994, 1995-1996), Tableaux de l'Economie Française.

INSEE (1995-a), 'Revenus et patrimoine des ménages', Synthèses, n°1, juin.

INSEE (1995-b), 'Comptes et indicateurs économiques, Rapport sur les Comptes de la Nation 1994', Economie générale, n° 109-110-111, juin.

Izraelewicz E. (1994), 'Les "vingt merveilleuses"', Le Monde, 17 décembre.

Jonas H. (1979, éd. fr. 1990), Le principe responsabilité, Une éthique pour la civilisation technologique, Paris, Ed. du Cerf.

Latouche S. (1986), Faut-il refuser le développement? , Paris, PUF.

Latouche S. (1994), 'Développement durable: un concept alibi, Main invisible et mainmise sur la nature', Revue Tiers-Monde, tome XXXV, n° 137, janvier-mars, pp. 77-94.

Latouche S. (1995), La mégamachine, Raison technoscientifique, Raison économique et mythe du progrès, Paris, La Découverte/M.A.U.S.S.

Maddison A. (1995), L'économie mondiale, Analyse et statistiques, Etudes du Centre de Développement de l'OCDE, Paris.

Meral P. (1995), 'L'équité intergénérationnelle en économie de l'environnement: réflexion critique de l'utilisation de la théorie de la justice de Rawls par l'analyse standard', Economie appliquée, 'Quelle économie pour l'environnement', tome XLVIII, n° 2, pp. 305-333.

Nikonoff J. et al. (1995), La République du droit à l'emploi pour chacun, Paris, Un Travail pour Chacun.

Norton BG. (1989), 'Intergenerational equity and environmental decisions: a model using Rawls' veil of ignorance', Ecological Economics, vol. 1, n° 2, mai, pp. 137-159.

Passet R. (1979), L'économique et le vivant, Paris, Payot, 2° éd. (1996), Paris, Economica.

Passet R. (1985), 'L'économie: des choses mortes au vivant", dans Encyclopaedia Universalis, Symposium, "Les enjeux', pp. 831-841.

Pearce D. (1987), 'Fondations of an ecological economics', Ecological Modelling, 38, pp. 9-18.

Polanyi K. (1944, éd. fr. 1983), La grande transformation, Aux origines politiques et économiques de notre temps, Paris, Gallimard.

Quarre D. (1995), 'Les salaires des agents de l'Etat en 1992', INSEE Résultats, n° 388-389, Emploi-Revenus, n° 84-85, avril.

Rawls J. (1971, éd. fr. 1987), Théorie de la justice, Paris, Seuil.

Sen A. (1993), Ethique et économie, Et autres essais, Paris, PUF.

Toman MA., Pezzey J., Krautkraemer J. (1994), 'Neoclassical economics and "sustainability"' Symposium international C3E, 'Modèles de développement soutenable, Des approches exclusives ou complémentaires de la soutenabilité? ', Paris, 16-18 mars 1994, vol. I, pp. 3-25.

Van Parijs P., 1991, Qu'est-ce qu'une société juste? Introduction à la pratique de la philosophie politique, Paris, Seuil.

Subject index

FSC
www.fsc.org
MIX
Papier | Fördert
gute Waldnutzung
FSC® C083411

Kolibri 360 GmbH
Ferdinand-Jühlke-Straße 7
99095 Erfurt, Deutschland
produktsicherheit@kolibri360.de